The Conservation Response

The Conservation Response

Strategies for the Design and Operation of Energy-Using Systems

Lloyd J. Dumas
Columbia University

Lexington Books
D.C. Heath and Company
Lexington, Massachusetts
Toronto

Grateful acknowledgment is made for permission to reprint an excerpt from the news report "U.S. Called Worst in the West in Conservation of Energy" *The New York Times,* November 28, 1975, © 1975 by The New York Times Company. Reprinted with the permission of United Press International and The New York Times Company.

Library of Congress Cataloging in Publication Data

Dumas, Lloyd J
The conservation response.

Includes index.
1. Energy conservation. 2. Energy consumption. I. Title.
TJ163.3.D85 621 75-10059
ISBN 0-669-08326-7

Second printing, September 1977.

Published simultaneously in Canada.

Printed in the United States of America.

International Standard Book Number: 0-669-08326-7

Library of Congress Catalog Card Number: 75-10059

Dedication

To the nations of OPEC who, in pursuit of political and economic power, may have finally kicked the United States hard enough in the seat of the pants to jar it into a rational energy policy—in the hope that the pain they have inflicted will not have been suffered in vain.

Contents

List of Tables

Preface

What with the lengthening gasoline lines and the predictions of a considerable shortfall in heating fuel during the coming winter, it was not at all difficult to become interested in the energy problem in the last months of 1973. It was at that time that my colleague, Professor Seymour Melman, suggested that we turn the attention of our departmental graduate Seminar in Engineering and Industrial Economics to the question of technical and economic alternatives for energy conservation. For each of the next three years, one semester of the seminar was devoted to that problem. The papers produced by the students in the Seminar, combined with the topics covered in a series of lectures delivered by energy experts from outside the University, soon made it clear that the potential for energy conservation was far greater than I had ever imagined. It became obvious that there was a strong case for energy conservation, and that there was a need for a broad, interconnected analysis of how that conservation might be achieved without damage to the standard of living.

In preparing this book, I have drawn heavily on the materials developed in conjunction with that Seminar. Some fifteen of the student papers prepared for the Seminar are specifically referenced in the course of the book, but in a sense I owe a much broader debt of gratitude to all those students and others who gave the Seminar life during those years.

My thanks to Diane Pagnott for her cheerful and efficient service in typing the manuscript. My special thanks to my colleagues Professors Seymour Melman, and Harold Elrod, and Roberto Agosti for their helpful comments; to Susan Straw for her encouragement and continuing interest; and especially to Andrea Dumas for her opinions, comments, encouragement, and, in particular, her patience, which helped me more than I can ever say.

Lloyd J. Dumas

Yonkers, New York
March 1976

Introduction

The abundance of energy has for some time now been considered a keystone of economic growth and increasing material prosperity. The industrial, mining, and agricultural sectors of developed nations consume prodigious amounts of energy in the production of the bewildering variety of goods that have come to be associated with modern economic activity. Vast and expanding quantities of energy are also consumed in the production of services, often considered the hallmark of an advanced economy, as well as in the use of the entire spectrum of goods and services by ultimate consumers. Without large quantitites of energy, modern economic societies would simply cease to exist.

It is therefore no surprise that when the leaders of some of the world's most important energy supplying nations drastically reduced that supply in the fall of 1973, enormous shock waves propagated through the economies of much of the developed world. The shock was more than economic–it was social and psychological as well. The economies of the developed world had been built on the assumption that abundant and relatively inexpensive energy would continue to be available into the indefinite future. Now that assumption was suddenly called into question–not by a Malthusian-style academic treatise–but by hard physical reality which could neither be debated nor ignored.

Nowhere did this shock more keenly penetrate the psyche than in the United States; for although it was among the nations least seriously threatened because of its sizeable domestic energy supplies, energy had always been so abundant and so cheap in the United States that many Americans had almost come to regard plentiful, inexpensive energy as a birthright. Inevitably, the first reaction was to begin the search for alternate supplies, particularly supplies not subject to arbitrary political interruption. By developing new supplies life could go on as before, after only a relatively short period of disruption. To be sure, exhortations to save energy by lowering thermostats and turning off lights in unoccupied rooms were made. But, the emphasis was clearly on expanding the energy supply–build the trans-Alaska pipeline, drill for offshore oil, develop better oil shale and tar sand extraction techniques, speed the construction of nuclear power plants, etc. Only secondarily, and somewhat reluctantly, was attention turned to the problem of energy conservation.

The phrase "energy conservation," like the term "budgeting," has a certain ring of asceticism to it, which does not accord well with the quest for the more abundant life. It conjures up images of working by candlelight, shivering in the cold, and bumping along the streets in a dirty, overcrowded bus. It seems particularly unpleasant to generations of Americans taught to believe that living in an appliance-packed, single-family house in the suburbs was the key to happiness and that the ability to drive endless miles in your own oversized, overstuffed automobile was the meaning of freedom.

But energy conservation, like budgeting, has much to recommend it. It is a way of ensuring that we will get the most out of available resources, no matter how scarce or how abundant they may be. Energy conservation is the path to maintaining or improving the standard of living in the face of limited energy resources—a path that is also compatible with the reduction of environmental pollution and the conservation of other natural resources.

This book is a systematic discussion of strategies for the conservation of energy. It is intended neither as an extremely specific reference handbook for the design of energy systems nor as a broad brush statement of the general problem. Rather, it is an area by area analysis of particular approaches to the design and operation of more efficient energy-using systems. Though this analysis is clearly not comprehensive, it nevertheless deals with a considerable variety of energy systems in some detail. A Summary and Conclusions section is included at the end of each chapter (except the last) so that the reader may easily sample its contents and delve more deeply in to the chapters or sections of particular interest. It is suggested, however, that even the reader with interest in less than the full range of energy conservation strategies read the first and last chapters in their entirety.

U.S. Called Worst in West In Conservation of Energy

PARIS, Nov. 27 (UPI)–The International Energy Agency said today that the United States was doing less than any other industrialized Western country to conserve energy.

The I.E.A., set up at Secretary of State Henry A. Kissinger's request a year ago to deal with the energy crisis, gave the United States the poorest rating among 17 of the West's leading industrialized countries.

"The American program must overcome an extremely high per capita historical energy consumption pattern and as such must be comprehensive and strong to be effective," an I.E.A. report said. "At the present time, it is neither."

The New York Times
November 28, 1975

1 Energy Conservation: the Possibilities, the Limits, the Benefits

Social Choice and the Energy Problem

The engineering design of any product begins with a central design objective and an explicit or implicit list of design criteria. The central design objective consists of the effective performance of the primary function for which the product is being created. The design criteria relate to characteristics of the product relevant to its acquisition and use and imply a set of subobjectives. For example, consider the design of a refrigerator. Here the central design objective is clearly the preservation of food through cooling. Some of the subobjectives implied by the design criteria might be: ability to cool rapidly, ability to maintain a constant temperature, high reliability, low first cost, low operating cost, high portability, low power consumption, low noise, high ratio of usable space to total size, high aesthetic appeal, etc.

No one design will be optimal for all criteria. There are inevitably tradeoffs in the design of a product insofar as achievement of the various subobjectives are concerned. For instance, a refrigerator designed with heavy emphasis on reliability may well have a higher first cost or a higher power consumption; greater portability may be achievable at the expense of slower cooling or increased noise. There are generally multiple tradeoffs in many directions. The key point is that many different designs are possible, all of which achieve the central design objective but differ in the degree to which various subobjectives are achieved. Thus, the establishment of the priorities attached to the relevant design criteria is of crucial importance to the characteristics of the ultimate product. This is universally true of all functional design situations, whether the designer is an engineer or an architect, whether the product is a toothbrush or an intracity transportation network.

This has two profound implications. The first is that *the level of technical knowledge extant at any given time does not dictate a single design for any given product.* Thus we are not faced with the binary choice of either accepting the existing design of a product or doing without. The second is that *the selector of the design criteria priorities plays a critical role in directing the application of technological knowledge.* What is crucial is that the selection of these priorities does not require any technical skill whatsoever. The establishment of priorities is not scientifically determinate. Rather, it is a matter of social choice.

What has this to do with the conservation of energy? In the past, particularly in the United States where energy was relatively cheap and its supply seemingly endless, the criteria related to energy consumption were assigned a very low priority in the design of most energy-using products and systems. Only in the few cases where energy costs were clearly a major part of total cost (e.g., in aluminum refining and electricity generation) was real weight attached to the reduction of fuel consumption. As a result, most energy-using systems were designed in such a way that energy consumption was greatly in excess of the minimum dictated by technical considerations related to the achievement of the primary design objective. This is as true of complex, multifaceted systems (such as the food processing system) as it is of the design of simple, single products. The result of the assignment of such low priority to the energy conservation criterion has been an almost unbelievably profligate consumption of energy. But, it need not be so.

By attaching a higher priority to the subobjective of minimizing energy use, products may be designed and systems developed that consume far less energy but still effectively achieve their primary design objective. Thus, through redesign and/or modification well within the scope of present technological knowledge (and frequently not involving significant life-cycle cost increase), energy consumption may be sharply curtailed *without* major sacrifice in the standard of living. There will of course be some costs, either financial or in terms of other criteria, such as convenience, since reduced energy consumption can only be achieved by tradeoff against the other criteria. But, surely we have all learned by now that "there is no such thing as a free lunch." However, as will be seen in subsequent chapters, in the case of energy conservation, the "lunch" will often be very cheap because of the nearly total neglect of the energy criterion in past design.

Since the setting of design criteria priorities plays such an important role in shaping the finished system, it is only reasonable to ask who sets these priorities and how this is accomplished. In the case of products offered for sale, the priorities are established by managements influenced preeminently by considerations of saleability. If "custom" styling and high-powered engines are more effective in generating automobile sales than safety features or gas economy, car designers will design flashy, high-powered gas guzzlers with little attention to safety. This means that the set of current and potential purchasers of the product, *taken as a whole,* play a critical role in the determination of design priorities. It is important to understand that it is the aggregate group of purchasers that influence design priorities, as a group, primarily through their purchasing patterns. Of course, whether a product is produced for sale or not, the priorities assigned to design criteria can also be influenced by direct governmental intervention as, for example, in the case of pollution control regulations placed on the automobile industry. Whether set by government decree or by customer purchasing patterns, the key point is that the establishment of design criteria priorities is a matter of social choice, *not* scientific necessity. Social choice set the pattern of priorities that resulted in the proliferation of devices with a voracious appetite for energy, and social choice can just as easily alter the pattern of priorities in directions that will curb that energy appetite.

The development of new technology is also of great importance to the conservation of energy. As will be pointed out in later chapters, there are many areas where technological progress could make a major contribution to energy conservation (e.g., in the improvement of energy storage devices). But technological development is not unidirectional, and, as in the case of design, the priorities followed play a major role in shaping the end result. In research, what is found is strongly conditioned by what is sought and where attention is directed. To be sure, not everthing is possible, and yet technology may be bent in many directions. Scientific progress does not follow a narrow single-lane road but rather a broad interconnected network. Which paths are explored and which are ignored is largely a matter of research priorities, and like design priorities, research priorities are also socially, not scientifically, determinate.

Now, apart from the excess consumption of energy arising from product and system engineering considerations, a sizeable fraction of the energy presently used is unnecessarily consumed, simply because of wasteful operating practices. Two examples so commonplace that all of us can undoubtedly verify them from personal observation are: unnecessarily prolonged idling of trucks, buses, taxis, cars, etc.; and space heating or air conditioning operated with windows and/or doors open. These are neither the most striking nor the most important examples, but they are among the most commonly observed. Such wasteful practices are extremely widespread, occuring even in areas where energy is quite expensive and uncertain in supply.[a] They appear to be largely the result of a combination of ignorance, laziness, and the fact that often the operators of energy-using equipment do not directly bear its energy cost (e.g., truck drivers) and/or that energy cost, though high, is a small fraction of total cost. In any case, the elimination of at least the more blatant of these practices can produce considerable energy savings without any negative effects on the standard of living. In the United States these practices are so commonplace that the potential energy savings are enormous.

In a sense, the modification of operating procedures so as to drastically reduce outright wastage of energy is both the cheapest and least painful energy conservation measure conceivable. Typically there are no capital costs required to

[a]For example, in the summer of 1975 the author observed numerous small shops in Japan operating air conditioners while the doors of the shops were wide open. (In some cases, no doors were present.) And this in a strongly energy-dependent nation, poor in energy resources, whose economy had been severly disrupted by the 1973–1975 OPEC oil price increases!

implement these changes, and operating costs are almost invariably reduced rather than increased. In addition, the lead times required for implementation are nearly always trivial. Frequently there are not even any intricate mental processes involved. It is largely a matter of breaking old energy-wasting habits and replacing them with habits that conserve energy.

Thus, by a combination of redesign, well within the bounds of current technical knowledge, and modification of operating procedures, considerable energy savings may be posted within a relatively short period of time. Redirection of technical research priorities along appropriate lines has great promise for generating large energy savings in the somewhat longer run. And all of these savings are attainable with minimal negative effect on the standard of living.

Energy and the Balance of Nature

Even if an infinite supply of energy resources were readily available in convenient form, the energy problem would not be solved. Attention would still have to be paid to the rate of energy consumption, because apart from the ecological problems associated with the extraction of energy resources, the consumption of energy itself has very important ecological effects. In fact, the ecological impact of consumption of energy resources is probably much larger than the impact of their extraction. For example, though the damage done by oil spills from offshore drilling operations, tanker accidents, and pipeline breaks is a serious matter, according to a 1972 report of the United States Geological Survey and the Environmental Protection Agency, more than two-thirds of the oil polluting the world's oceans comes from the disposal of waste oil from automobile engines and other machines.[1]

The ecological problems directly resulting from energy consumption can be divided into two broad classes: the pollution associated with particular modes of consumption and/or particular energy resources (e.g., the burning of gasoline in automobiles) and the heat pollution inherent in any use of energy, no matter what the source. The importance of this differentiation is that the former class may be largely eliminated, at least theoretically, by use of completely "clean" fuels like solar or tidal energy, but the latter class cannot. This is not to say that the thermal pollution problem cannot be greatly mitigated by the improved design and operation of energy-using systems, for it certainly can be. But, it is to say that there is something more inevitable, more fundamental about the thermal pollution consequences of energy use than those of any other type of pollution.

Pollution is defined in essentially biological terms. That is, pollution is something that alters the environment in a way that is harmful to life forms.

By implication, whether something is considered a pollutant or not depends on the perspective of the one doing the defining, in the sense that focusing attention on different life forms yields a different set of pollutants. For example, the residue from phosphate detergents was not a pollutant to the algae that thrived on it but certainly would be to human beings. There are, of course, common requirements of essentially all forms of life as we know it on earth (e.g., oxygen, warmth), and anything that operates in any way to deprive living organisms of these needs is a pollutant. But the diversity of life is such that this set of universal pollutants is not as large as one might at first think.

Since it is human beings who write books and give speeches about pollution, the discussion progresses with a completely understandable bias in favor of our own species and those life forms that we consider congenial. There is nothing in particular wrong with this. Unfortunately, this tendency has often predisposed us to either purposely damage those life forms that we considered bad or ignore the adverse consequences of our actions of life forms not at the top of our list, without paying sufficient heed to the impact of what we have done on the ecological web. It is only recently that we have begun to understand that the intricate feedback mechanisms of the ecological net will, in due course, cause us to pay for the damage we do. We are learning, though perhaps too slowly, that the ecosphere is not to be trifled with.

The ecosphere, of course, consists not only of the other life forms with which we interact but also of the physical environment. Energy consumption has its primary adverse ecological impact in the contamination of the physical environment. The pollutants released to the environment as a consequence of our energy use affect us in two ways. To the extent that these substances infiltrate and damage our bodies, we are directly affected. But even those substances that either do not enter our bodies or are not directly harmful to them may still indirectly cause us serious problems if they damage or favor other life forms, disrupting the ecological balance. Thus it is to our advantage to seek an understanding of the ecological effects of all substances we release to the environment in hypernatural concentrations, whether or not they have obvious direct effects.

Having long ago observed that even extremely harmful substances could be tolerated with little or no effect if they were present in low enough concentrations, our whole philosophy of disposal of pollutants has primarily been one of dispersal. We have operated under the assumption that, as G. Tyler Miller has put it, "dilution is the best solution to pollution."[2] However, there are four reasons why this approach is not indefinitely workable: bioconcentration, synergistic interactions of pollutants, the problem of local damage, and the limits of environmental absorption.

The natural, in a sense entropic, tendency of pollutants to disperse when they are dumped into the water or the air can be undone by the process of bioconcentration. For example, some of the harmful hydrocarbons found in the oil which pollutes the sea are absorbed or ingested by marine organisms. As these organisms become food for organisms higher up in the food chain, these hydrocarbons may be progressively more concentrated in the tissues of each successive organism. Eventual ingestion of certain types of seafood by people, directly or through the intermediary of livestock fed fish meal, can then provide doses of the toxic substance in concentrations far greater than that found in the seawater. There is some evidence that among the chemicals that may be concentrated in this way are well-known carcinogens.[3] Any substance that is neither voided nor chemically altered by the organisms that ingest or absorb it is a candidate for bioconcentration.[b] In effect, the organisms lower in the food chain serve as gatherers of the substance, collecting it for those progressively higher up. In this way, the dilution which occurs as a result of physical processes may be reversed by biological processes.

Pollutants that are too dilute to cause trouble by themselves may interact in ways that can potentially produce serious problems. Apart from the possibility of direct chemical reactions between pollutants and interactions of pollutants with the physical environment (e.g., photochemical smog), each toxic substance a plant or animal encounters constitutes another assault on the organism. Even if each substance interfered with a different part of the life process, each assault may render the organism more vulnerable to every other assault. Thus, even where there is no dangerous chemical or physical synergy, there may well be a disastrous biological synergy.

The local damage problem is generated by the fact that pollutants released to the environment are not diluted to the eventual equilibrium levels immediately. Rather, the concentrations of pollutants in both space and time near the source of pollution will tend to be high. Consequently, the local environment may well be subject to levels of contamination far in excess of the average. In many cases, these levels will also be in excess of safe levels, particularly for prolonged exposure. The local environment may thus be rendered hostile to some life forms and/or perhaps extremely favorable to other life forms dangerous to humans. To the extent that there are concentrations of such life forms in the vicinity, the biological component of the local ecology may be severly damaged. This, either directly or indirectly, can readily produce effects deleterious to people. Local pollution can also produce far-ranging effects through the alteration of the physical component of the local ecology; for example, through the modification of wind or rainfall patterns. Even though these changes may not seem important or dangerous because they are small, it is worth bearing in mind that in any feedback network as intricate as the ecosystem, major changes may be brought about by relatively minor disruptions that exceed certain critical triggering levels. What is particularly troublesome is that, in most cases, we do

[b]The process of bioconcentration can be shortcircuited, at least at the upper levels of the food chain, if the substance is concentrated in a part of the body not considered edible by the predator on the next higher level.

not even know what these triggering levels are or precisely how such triggers would operate. But there is no question that they exist. Anyone who has ever been present when a sound was made into a live microphone inadvertently placed in front of its own speaker knows that the magnifying capabilities of feedback systems are impressive, and often quite distrubing.

The issue of limits of absorption is closely related to the existence of "ecotriggers" and to the local damage and synergy problems as well. The environment simply cannot continue to absorb and dilute pollutants indefinitely without undergoing serious changes. For example, consider the case of thermal pollution. Since heat is invariably generated by processes involving the consumption of energy, the ultimate limiting factor on energy consumption is the capacity of the environment to absorb and dissipate heat *without seriously upsetting the ecological balance.* But because of ecotriggers, local concentration of energy consumption, and the synergistic effects of the generation of heat and other environment-changing human activities, the operational heat tolerance of the ecosphere is not nearly as great as one might assume.

Most of the solar energy passing through the atmosphere is in the form of visible light, but much of what is reradiated is in the infrared part of the spectrum. While the atmosphere is relatively transparent to visible light, it is relatively opaque to infrared. The result is that the air acts much like the glass in a greenhouse, warming the lower atmosphere and the surface of the earth. The levels of carbon dioxide and water vapor in the air are critical to the strength of this so-called "greenhouse effect." Now, the combustion of fossil fuels is responsible for the addition of large amounts of carbon dioxide to the atmosphere (estimated to rise to 26.0×10^{12} kilograms in the year 1980 alone[4], thus strengthening the greenhouse effect. As the warming proceeds, there is a tendency for absolute humidity to increase due to the increased ability of the atmosphere to retain moisture. But the additional water vapor absorbed by the air adds to its opacity to infrared, further augmenting the greenhouse effect. Clearly, the larger the quantity of heat trapped via the greenhouse effect, the more severely the ecosphere's capacity to tolerate additional heat is taxed.

Apart from these global effects, the pattern of human energy consumption activities also contributes to altering the ecological heat balance by generating differential local effects. Cities have become islands of pollution. Heat rises from them as a result of concentrated energy consumption activities, as do clouds of gaseous and particulate pollutants. These effects, along with changes in albedo and wind flow due to buildings and pavement, produce alterations in the local climate. Cities tend to be cloudier, have more precipitation, less humidity, higher temperatures, and lower winds than surrounding rural areas.[5]

Well before human energy consumption activities change the mean global termperature the few degrees needed to trigger the global processes that can create havoc in the physical component of the environment (e.g., melting of the polar ice caps), their contribution to climate modification may produce disastrous effects in the biological component. As Carol and John Steinhart have reminded us,

> A slight but persistent change in temperature or humidity can tip the balance in favor of organisms pathogenic to plants, animals, or man, causing severe epidemics of diseases and pests that formerly were held in check by climatic conditions. Major epidemics of crops and forest trees in the past have often been associated with just such changes. Or the major winds could shift their course, and fail to bring rain at the usual season; and millions of people who depend on the rains to water their crops could starve. The monsoons have not been visiting some regions as reliably as they once did, and some meterologists are building a convincing case against air pollution as the ultimate cause[c]. . . .[6]

Thus, the ecological limits of absorption are much more restrictive than the physical limits.

Pollution Control and the Energy Crisis

Even before the oil embargo of 1973 spectacularly catapulted the energy problem into the limelight, the groundwork for the argument that there is an essential conflict between the achievement of a clean environment and the solution of our energy problems had been established. For years the energy-supplying industries had battled with the environmentalists on the issues of strip mining of coal, off-shore oil exploration, the trans-Alaska pipeline, the construction of nuclear power plants, etc. When the oil boycott changed the energy problem into an energy crisis, it was a simple matter for the energy industries to point an accusing finger at the environmentalists and cite their ecological victories of the 1960s as a major source of the national dilemma. After all, was it not the environmentalists who had blocked the expansion of domestic oil supplies, coal production, and the increased generation of electricity by nuclear fission? Was it not they who had insisted on the use of only low sulfur fuel oils in urban areas and pressed for the installation of gas wasting pollution control devices on automobiles? A casual glance at the simultaneously rising level of Gross National Product (GNP) and energy consumption over the long term showed that continued economic prosperity depended on continually expanding energy consumption, and so by hamstringing our efforts to increase domestic energy supplies and pressing us into fuel-wasting pollution control measures, the environmentalists had threatened the nation's economic future. If that were

---

[c]The largest portion of most types of air pollutants are injected into the air as the result of disposal of the waste products of fuel combustion. For example, in 1968 in the U.S. the burning of fuel for transportation, power generation, home heating, etc. was responsible for 65.7 percent of the carbon monoxide, 75.9 percent of the sulfur oxides, 54.1 percent of the hydrocarbons and 87.8 percent of the nitrogen oxides polluting the atmosphere, according to J. Holdren and P. Herrera, *Energy* (San Francisco: Sierra Club, 1971), p. 145.

not enough, the oil boycott made it clear that the increasing inability to supply the nation's energy needs from domestic sources had also compromised our future in international politics and laid us open to blackmail. Thus as the argument goes, the so-called ecological triumphs of the 1960s had produced economic and political disaster.

Under this onslaught, the initiative began to pass into the hands of energy industry, and environmentalists seemed to be put increasingly on the defensive. The stunning success of the opponents of the trans-Alaska pipeline was dramatically reversed. Increasingly, pressures were applied for delay or relaxation of pollution emission standards and for expansion of off-shore drilling, construction of nuclear electric facilities, development of oil shale deposits, etc. The whole range of achievements of the ecology movement in the 1960s is under attack, and a public that was willing to put up with some inconvenience and added expense in the name of a cleaner environment has shown itself patently unwilling to tolerate major economic dislocation.

Fortunately, we do not have to choose between economic and ecological well-being, for the argument that to solutions to the pollution and energy problems are necessarily antagonistic is based on a combination of a misreading of history and a misunderstanding of technical facts. These problems are susceptible of simultaneous solution—through the conservation response.

By improving the technical efficiency of energy-using processes, essentially the same standard of living may be maintained with a smaller input of energy resources. This clearly reduces the pressure against energy supplies. And it also tends to reduce the level of pollution, since even if the same amount of pollution is generated per unit of fuel consumed, there will be fewer units consumed. However, to the extent that they allow more complete use of the fuels involved, modifications that lead to higher energy efficiency may also reduce the level of pollution, even on a per unit basis.

For example, consider the case of fossil fuels. Fossil fuels are burned to supply energy for many purposes including transportation, space heating, water heating, and electricity generation. Their combustion is also responsible for the lion's share of air pollution.[7] Now, fossil fuels are essentially made up of hydrocarbons, and complete combustion of a pure hydrocarbon produces only carbon dioxide and water as by-products to the generation of heat, neither one of which is particularly toxic. On the other hand, incomplete combustion yields carbon monoxide along with particulates and some unburned hydrocarbons, all of which are dangerous pollutants. Thus more complete combustion of fossil fuels will simultaneously produce more energy and less pollution per unit of fuel burned. Likewise more complete use of the heat generated in the combustion process will result in less waste heat being produced and, therefore, less thermal pollution per unit. If more useful energy is generated per unit of fuel used, fewer units will be needed to accomplish a given result. But this, along with a reduction in pollutants emitted per unit of fuel consumed, implies a considerable reduction in the total amount of pollution associated with the combustion. Therefore improvements in the technical efficiency of fossil fuel combustion not only save fuel but also reduce emission of gaseous and particulate pollutants.

Unburned hydrocarbons, carbon monoxide, and particulates are not the only air pollutants that typically accompany the burning of fossil fuel. Nitrogen oxides are also generated when the combustion takes place with air used as the source of oxygen. Sulfur oxides may also be generated where the hydrocarbon mixture is contaminated with sulfur, as it nearly always is in the case of oil and coal. The nitrogen and sulfur oxide problems are not solvable simply by making the combustion more complete, because they are caused by contamination, as it were, of the oxygen supply on the one hand and the fuel supply on the other. The reduction of these pollutants requires either removal of the contaminant before combustion or removal of the oxides from the output gases after combustion. Some approaches to this problem require additional energy or preclude the use of certain energy resources, implying that some degree of energy-pollution tradeoff exists. However, this tradeoff is by no means inevitable, even here. Solutions to some other ecological problems can work in the direction of undoing these difficulties. For instance, combustion or pyrolysis of agricultural and urban wastes not only mitigates solid waste disposal problems but also provides a large source of low sulfur fuel.

Recycling, an activity strongly associated with the ecology movement, is also a superb energy conservation measure. Over a whole range of products, the energy required to produce the good from virgin raw materials is vastly greater than the energy required for recycling. This is true whether the recycling procedures require only cleaning or reduction to scrap and reprocessing. For example, using refillable bottles would save about two-thirds of the energy required for nonreturnables[8]; production of aluminum ingot from scrap requires only about 5 percent of the energy required to produce it from ore[9]; and pulp made from recycled waste requires only one-fourth the steam and one-tenth the electrical energy of that made from wood.[10] Thus the recycling approach simultaneously reduces solid waste pollution and energy consumption.

Finally, it is also true that the only essentially inexhaustible energy resources are those which if properly used, generate the least possible pollution. Without question, solar energy is potentially the most important of these. It is showered on the earth in an abundant, well-distributed supply that does not lend itself to monopolization and can be relied upon into the indefinite future. In addition, wind, a somewhat less reliable form of solar-generated energy, tidal and other hydrogravitational energy (e.g., waterfalls), geothermal energy (e.g., hot rocks and hot springs), and hydrothermal energy (e.g., ocean temperature gradients) are all natural energy sources that neither pollute the air through combustion nor produce other by-products dangerous to the biosphere. Developing these sources properly will result in a greatly increased energy supply without additional pollution. In fact, replacing fossil fuels and nuclear fission energy by one or more of these forms will result in a drastic reduction of all gaseous and particulate pollution. However, even they will generate thermal pollution, which we have seen is an inevitable consequence of energy use. In addition, care must be taken to develop these sources in such a way that the means employed for tapping them do not produce undue ecological disruption, e.g., that we do not overuse ocean temperature gradients and thus disrupt hydrospheric circulation patterns with possibly disastrous climatic effects. These nonpolluting energy resources have great potential, and there is no technical reason why they cannot be exploited without serious ecological effects, *but* their use too is subject to the inherent limits imposed by the laws of physics and the balance of nature. This must not be forgotten. Therefore, attention to energy conservation will still be a wise policy, even after these sources are developed. Prior to that time, energy conservation can serve as a critical means of buying time for that development.

Though there are some places where a tradeoff exists between ecological and energy concerns, at base the solutions to the problems of energy and pollution control are largely congruous. The conservation response will buy time for the development of nonpolluting natural energy sources. It calls for improvements in the technical efficiency of energy-using systems which tend to reduce pollution as well as fuel consumption. Recycling, a policy strongly favored on energy conservation grounds alone, also has extremely favorable ecological implications. Finally, whatever the energy source, energy conservation acts to put distance between the ultimate ecological limits on energy consumption and the levels required for a high material standard of living. Thus the compatibility of the solutions to our ecological and energy problems is far more basic and more striking than their conflict.

The Mythology of Beneficent Waste

From both sides of the political spectrum one often hears the argument that waste is an absolute necessity to the proper functioning of the capitalist economic system. If products are built to fall apart or to drastically lose their appeal in a short time, that is good because only in this way will the productive system face the constantly renewed demand required to maintain its operation. If resources are poured into the production of economically useless goods,[d] that is good too because resources are employed that would otherwise be a glut on the market and the payments to the human resources involved provide them with

[d]Military goods are the classic example of goods that are economically useless. That is not to say they have no value, that is only to say they have no *economic* value. They contribute nothing to the current standard of living and nothing to the economy's capacity to produce economically useful goods in the future. Thus resources poured into military production are, from the purely economic point of view, wasted. For an interesting and detailed analysis of the economic effects of military spending see S. Melman, *The Permanent War Economy* (New York: Simon & Shuster, 1974)

additional purchasing power to demand more goods and thus further stimulate the productive system. Finally, if the products themselves must be built so that they will rapidly be thrown away, what conceivable sense does it make to be overly concerned with the wastage of fuels and other resources in the production of those goods?

The capitalist system, so the argument goes, not only produces waste—it thrives on it. Conservation of any sort is anathema to be tolerated only when a particular vital resource is running short and only until our technical ingenuity can fix us up, either by locating new supplies or by finding substitute resources. Any permanent, long-term commitment to conservation is sure to bring on economic catastrophe.

The idea that waste is beneficial to the capitalist, or any other, economic system is one of the most widespread, well-accepted, and tenacious of the socioeconomic myths—but it is a myth nevertheless. In economy, as in ecology, waste is sometimes tolerable, but it is never beneficial.

In order to understand why waste is not necessary to the proper functioning of the economic system, let us postulate an extreme hypothetical case of a productive system with the following characteristics: (1) no economically useless goods are produced; (2) there is no "planned obsolescence," i.e., no goods are designed to fall apart or go out of style rapidly; (3) no "scrap" is generated in the production of any good; and (4) no good produced ever physically wears out. It should be noted that only the first two characteristics are even theoretically attainable. Nevertheless, all of them are postulated in the interest of constructing a maximally wasteless economy.

The economic system will be said to be capable of functioning properly if it is capable of operating so as to produce as high a material standard of living as possible, given the amount of resources available. People will be assumed to engage in work, as opposed to any other activity, only for the purpose of acquiring the goods and services produced within the economy. Work is defined in this way to order to emphasize that the other benefits, if any, that people derive from the typically social activity we normally call work can, in principle be derived from other nonremunerative activities.[e]

First, consider the functioning of our hypothetical economy from the firm's viewpoint. If the firm is the producer of a service or any good that is consumed (e.g., food, cleaning fluids, paper, ink) it can operate in essentially the same fashion as present-day "real world" firms do. The fact that its customers use up the product assures that it will face a renewable demand and thus will continue to purchase raw materials, use energy, and employ labor in the pursuit of profits or sales. But what of the firm that produces one of the everlasting goods we have postulated?

---

[e]That is to say, human activity is not classified into the typical economic work-leisure dichotomy. Rather, work is only taken to be one of a full spectrum of human activities.

Suppose a new durable good called "television" is invented and that a firm is formed for the purpose of exploiting the profit potential of this new product. The firm acquires machinery, buys materials, hires labor, and begins producing televisions. Initially, no one owns a television, so assuming that people either want the product or can be convinced to do so, the potential market is large. But as more and more televisions are sold, the market eventually becomes saturated and sales and profits tend to fall off. Since by assumption the televisions never wear out, there is no replacement demand. If there is population growth, there may be some small residual demand,[f] but let us assume this away. What future is left to the firm besides bankruptcy and dissolution?

In the first place, any significant technical improvements in the product may produce a strong new spurt of demand, e.g., the development of portable TV or color TV. In the second place, the firm may shift its production to a totally new durable product, whose demand has not yet been saturated, or possibly even to the production of a service or nondurable good. But there is a still more fundamental way to look at this situation.

Even if there is no possibility for the firm to convert its production, so that it must go out of business, is there anything economically troublesome about this? Put differently, the essential question is: do profit opportunities have to be eternal in order to be exploited without dire economic consequences? If you owned the only oil deposit on earth, would the fact the removal of the oil would eventually deplete it prevent you from developing the deposit, even though you knew no other deposits existed? Would the fact that that activity, and the gain associated with it, was only temporary necessarily cause you or anyone else great hardship? The answer to both these latter questions is obviously no. Just as your attention could be turned elsewhere after the well ran dry, so can labor resources in general be turned to other forms of useful activity when one form is terminated. Often the same is true to a lesser extent of nonhuman resources as well. To be sure, the transition will not automatically be smooth and relatively painless in all cases, but careful attention to the resource conversion problem can eliminate any severe economic difficulties.

From the workers' point of view, our hypothetical, wasteless economy not only presents no serious difficulties, it actually has some considerable advantages. Because durable goods never wear out, people can obtain the services of these goods indefinitely once they have been purchased without ever having to worry about replacing them. They need only cover their operating expenses (e.g., electric power costs). Technically, these goods could be handed down from generation to generation. However, people would still need to work to but services and nondurable goods. If they so desired, they could also work to enable

[f]Since the production time for any good tends to be very short relative to the production time for a new household, the continued demand from any population growth will be very anemic.

themselves to increase their stock of durable goods or to acquire newly developed goods or technologically superior versions of previously existing goods. Thus, if the hypothecated eternality of durable goods implies somewhat less labor demanded for the production of durable goods, it also implies less work required for the acquisition of durable goods.

It is therefore clear that even an economy based on essentially capitalist market-profit motive principles does not require waste to function, and that is the central point of our temporary flight into fancy. But, in fact, the case is much stronger than that. Waste in the productive system represents opportunities foregone.

Since all resources, including labor, are limited in supply, every waste of fuel or raw materials and every bit of human effort wasted in the production of useless goods or replacements for goods built to fall apart prematurely must be considered as imposing a cost. The cost may be reckoned in terms of the useful goods and services that could have been produced with the resources that were wasted.

A serious commitment to energy conservation, therefore, not only makes ecological sense but conveys economic benefits as well. Waste of energy or any other productive resource, far from being a sign of prosperity and economic well-being, is a sign of profligacy and myopia. It is a parasite on the productive system which we may not be able to eliminate but can certainly hold in check.

The Thermodynamic Underpinnings of Energy Conservation

Strictly speaking, energy is not consumed in the sense of being used up in the everyday world of chemical and nonnuclear physical processes. The so-called "law of conservation of energy" states that energy can neither be created nor destroyed by such processes but rather only transformed from one form to another. The forms of energy include light, heat, mechanical, electrical, and chemical energy.

One of the most important energy relationships for our purposes is that between the flow of heat and the performance of work. *Heat* may be defined as energy that flows between two bodies because of a temperature differential, while *work* is broadly defined to include all other energy flows, i.e., work is energy transmitted from one system to another without a temperature difference being directly involved. The branch of physics that studies the heat-work interactions of systems and the changes of energy involved in the performance of work and the flow of heat is called *thermodynamics.* The law of conservation of energy stated above is also called the *first law of thermodynamics* and may be readily reformulated in terms of the heat-work dichotomy as any change in the internal energy of a system must be exactly equal to the difference between the heat flowing into (from) the system and the work being done by (on) the system. Thus no energy is, so to speak, lost.

Thermodynamics differentiates between two types of processes—those that are *reversible* and those that are *irreversible.* Since reversible processes exist in an idealized form (such as the Carnot cycle) while essentially all natural processes are irreversible, we will focus on the latter. An *irreversible* process involves the change of an isolated system from state A to state B such that the system cannot return to state A without outside intervention. In other words, an irreversible process is not irreversible in the sense that it *cannot* be undone, it is irreversible in the sense that it *will not* be undone by the system itself. For example, suppose the system consists of a hot body and a cold body in contact with each other but insulated from the external world, i.e., from the environment. Clearly, heat will flow from the hot body to the cold body until they have both reached the same temperature. If we call the hot body-cold body initial situation state A and the warm body-warm body situation state B, then the system has, by itself, moved from state A to state B. However, the system will not return to state A, i.e., the hot body-cold body state, without external interference.

The generalization of the empirical observation that systems behave in this way is called the *second law of thermodynamics* and can be stated as follows: no self-contained system can transfer heat continuously from a body at a lower temperature to a body at a higher termperature. An alternate statement of the second law is: heat cannot be extracted from a source in a self-contained, isolated system that is the same temperature throughout and transformed into work. The importance of temperature differentials to the ability of a heat engine to perform work is illustrated by the fact that the formula for the ideal efficiency of a heat engine is $e = (T_1 - T_2)/T_1$, where T_1 is the intake and T_2 the exhaust temperature (in degrees Kelvin). The net effect of the second law of thermodynamics is that heat cannot be transformed into work *unless* a temperature difference exists and that heat always naturally flows in the direction that tends to *eliminate* any temperature difference. Thus equilibrating heat flows are irreversible processes that involve a loss of the system's capacity to do work.

Pressure differentials are analogous to temperature differentials both insofar as the ability to perform work and the unidirectional tendency toward equilibration are concerned. For example, suppose a container of gas is placed inside a larger enclosure which contains a vacuum. If the gas is not prevented from doing so, it will expand to fill the enclosure. The pressure differential between the gas in the container and the zero pressure of the evacuated enclosure will result in a flow of gas from the container to the enclosure that will tend to eliminate the initial pressure differential. As in the hot body-cold body case, this too is an irreversible process. The gas will not flow back into the container and out of the rest of the enclosure unless some external influence interferes. Additionally, in the process of flowing from the container to the enclosure, the gas could have been made to do work, say by interposing a piston between the container and the rest of the enclosure. However, once the pressure differential has been eliminated, the system's capacity to do work disappears.

Since all natural processes that tend toward equilibration will not move back into a nonequilibrium state without external intervention, it is clear that such processes are irreversible. It is also clear that since such irreversible processes involve the elimination of temperature, pressure, etc. differences within systems left to themselves, they produce a loss in the capacity of those systems to perform work. Once the differences have been eliminated without any mechanical work having been extracted in the process, an opportunity for extracting useful work has been permanently lost.

Irreversible processes occurring wholly within an isolated, self-contained system will reduce its capacity to perform work, while reversible processes will leave the system's work capacity unchanged. Only by the interference of influences external to the system can its ability to perform work be increased. Since for the universe as a whole there are no external influences, and since all natural processes are irreversible, the work capacity of the universe must always be declining. Physicists have developed the concept of a measurable physical quantity, called *entropy,* which increases as the capacity to perform work declines. Thus the aggregate reduction of work capacity resulting from the elimination of temperature, pressure, etc. differentials can be said to give rise to increases in entropy for the universe as a system.

The universal tendency toward equilibration and hence declining work capacity may be somewhat philosophically disturbing, but it is nevertheless real and unalterable. However, it is clearly in our interests to minimize the decline in work capacity of the systems with which we directly deal (beyond that which is necessitated by our extraction of useful work), and there is no law of nature which prevents us from doing that. To the extent that we can avoid excessive dissipation of work capacity, we can minimize the necessity for boosting the system in order to extract further useful work. Since external boosting inevitably requires the injection of "new" energy into the system, it tends to increase the rate of consumption of energy resources associated with the performance of useful work, which clearly works in a direction counter to our interests. Thus minimization of the excessive decline in work capacity is an important objective in the design and operation of energy-using systems.

It is highly improbable that random molecular and energy processes will generate significant temperature, pressure, etc. differentials. On the contrary, such processes tend with very high probability to produce essentially even distributions. Thus the tendency of natural processes to equilibration can be seen as a tendency toward randomness or disorder. We have seen that this irreversible tendency of natural processes toward equilibrium involves a reduction in the associated system's capacity for work. Therefore increasing disorder is associated with declining work capacity.

Since heat is the most disordered from of energy (in fact, heat *is* disordered energy[11]), one operational implication of the above-stated objective of minimizing excessive decline in work capacity is simply that the unnecessary generation of heat is to be avoided. This is particularly true of low temperature heat or, more properly, heat flows that are associated with low temperatures. It is true because, as we have seen, the ability to extract work from a heat source is a function of the temperature differential between that source and an available lower temperature heat reservoir. The lower the temperature of the heat source the more difficult it is for us to produce a still lower temperature reservoir without performing additional work, i.e., without requiring use of additional external energy.

Thus the underlying strategy for the conservation of energy in the design and operation of energy-using systems is the minimization of the excess decline in the work capacity of the systems with which we deal. Operationally this simply means the avoidance of all avoidable dissipation of any form of energy. More specifically, it implies the minimization of the unnecessary generation of heat, particularly heat associated with low temperatures. Where the generation of heat is an essential part of the motive process, e.g., in the conversion of chemical energy to heat energy by combustion of a chemical fuel, this heat should be produced at the highest possible temperature. Following this strategy will result in the maximization of our capacity to extract useful work from our existing energy resources.

Summary and Conclusions

We have seen that much of the widespread wastage of energy in the United States and elsewhere is the result of conscious or unconscious social choice and not technical or economic necessity. Accordingly, this waste can also be terminated by a social decision to do so. The design of energy-using systems is affected by the priorities we attach to the various design criteria with which engineers and architects operate. Though few of us ever formally draw up an explicit list of weighted design criteria, our preferences, whether they be for style over safety or convenience over energy use, are communicated to the design process nevertheless. Since many designs are possible for any given product, a raising of the priority attached to limiting energy consumption can result in large gains in energy conservation with little or no sacrifice in the performance of the product's essential function. Similar considerations also apply to the mode of operation of energy systems. The ultimate conclusion is that large energy savings may be achieved with little sacrifice in the material standard of living.

We have also seen that there are essential thermodynamic limits on our ability to recycle energy and thus improve the efficiency of energy-using-systems. In the course of extracting useful work from an energy source, a portion of that energy is inevitably degraded. The second law of thermodynamics guarantees that irreversible natural process will waste some of the energy used, i.e., render it incapable of performing further work. But apart from indicating that there are physical limits on our ability to consume energy without generating waste, the laws of thermodynamics also suggest two fundamental conservation strategies in the design and operation of energy-using systems: (1) minimize the generation of all heat not directly required and (2) for equivalent amounts of waste energy, give preference to high temperature heat over low temperature heat.

The thermodynamic facts of life teach us that, at the very least, our consumption of energy must generate pollution in the form of heat. Of course, for a variety of reasons human energy consumption activities typically generate many other forms of pollution as well. Since these activities, of necessity, have impact, we are obligated to modify them is such a way as to minimize the ecological disruption they cause. The intricate ecological feedback system guarantees that we cannot and will not escape the consequences of any damage we do. Furthermore, the existence of ecotriggers, synergies, and local damage effects indicate that the level of pollution required to produce ecological disaster is far lower than we might, at first, expect. Thus the ecological constraints within which our energy consumption activities proceed are considerably tighter than the purely physical constraints.

Fortunately, the same logic that implies that we cannot consume energy without polluting also implies that we can attack the ecological and energy problems simultaneously. Energy conservation measures, such as the improvement of the technical efficiency of energy systems, also tend to reduce environmental pollution. Measures originally put forth as ecological policies, such as recycling, often tend to save potentially enormous amounts of energy. It even turns out to be true that the indefinitely available, renewable energy resources are precisely those that tend to be maximally clean, e.g., solar and tidal energy. Therefore, far from being inevitably antagonistic as has been frequently and vociferously alleged, there are solutions to the energy and pollution problems that are quite compatible.

Finally, we have tried to lay to rest the argument that energy conservation is economically dangerous because the economy requires waste to prosper. This argument is not simply misleading, it is absolutely wrong. The waste of any productive resource, including energy, imposes a cost on all of us. That cost is best measured in terms of those positive things that could have been achieved by the productive use of the resources squandered.

Energy conservation is therefore not only technically possible but also highly beneficial both economically and ecologically. Having attested to its feasibility and desireability, we now turn to the business of how it may be achieved.

Part I
Buildings

2 Building Design and Energy Consumption

The Architecture–Energy Interface

As of the early 1970s, somewhat more than one-fifth of all the energy consumed in the United States was used for space heating, air conditioning, or lighting in the residential and commercial sectors alone.[1] Although the data that would allow such a calculation is not readily available, it is clear that the total energy consumed by buildings, across all sectors, is easily more than 25 percent (possibly much more) of the total United States energy demand.[a] Thus buildings are a major point of energy consumption.

The single most important determinate of the life-cycle energy performance of any building is its architectural design. To be sure, operating procedures can do much to alter the pattern of energy consumption in the provision of building services, but they must always be carried out within the basic constraints established by the building's design. There is, or course, an obvious asymmetry here. While a building that is well-designed from an energy standpoint can be rendered inefficient by poor operating procedures, one that is poorly designed cannot be made efficient merely by improving those procedures.

The architectural process exerts influence on the energy requirements of a building in a number of ways. The choice of materials influences both the amount of energy built into the structure and certain characteristics of the finished building that have important energy-consumption implications. The selection of materials that are manufactured by more energy-intensive processes or that must be transported over longer distances increases the initial energy cost of the building. The thermal conductivity and thermal mass of the materials chosen, among other energy-relevant characteristics, influence the operating energy costs—costs that must be borne for long periods of time.

The structural design also impacts on both the construction and operational energy requirements. The former effect results from the interaction of structural design with the choice of materials in determining the quantity of materials required; while the latter effect is primarily due to the spacial relationships established, along with the properties of the structure relevant to thermal, optical, etc. considerations. The design of the service systems—those systems that provide heating, cooling, ventilation, and internal transport of all kinds—clearly has

[a]Here we mean to include only the energy required for constructing buildings and for maintaining a comfortable occupancy environment within them.

great energy significance, though the primary impact here is on post construction energy use. This is similarly true of the building's exterior design and orientation. Thus virtually all components of the architectural process exert influence on the energy requirements, or a building singly and by virtue of their interactions.

The extent to which buildings meeting essentially the same performance requirements can differ in the energy they require is perhaps best illustrated by reference to a few specific examples. A New York City based architectural firm's comparison of two office buildings in Albany, New York revealed that one of the buildings consumed 50 percent more energy per square foot than the other.[2] In a detailed study of the energy performance of the nearly one thousand buildings in the New York City public school system, architects Richard and Carl Stein found that the highest energy-using building consumed about two and one-half times the energy per square foot of the average of all the other buildings.[3] Of 86 office buildings built after 1945 in New York City responding to a survey conducted by the Municipal Services Administration, the highest energy-user consumed five and one-half times the energy per square foot consumed by the lowest energy-user.[4] It is important to emphasize that each of these comparisons involved buildings with essentially similar performance requirements, and that in no case was the higher energy consumption associated with demonstrably superior performance by the building occupants.[b] Furthermore, the bulk of the energy consumption differentials was apparently assignable to architectural design considerations.[5]

Architectural Philosophy and Energy Conservation

The natural environment provides the basic context within which all human activity takes place. Many of the essential environmental requirements of human beings are, at least to some extent, provided by that environment without human intervention. If that had not been the case, our species would never have been biologically successful on this planet. In a sense then, there is a certain core package of environmental services that are given to use essentially free i.e., without either the expenditure of human effort or the consumption of any of the earth's stock of depletable resources.

It is only when we wish to cause our local environment to differ from that which is naturally provided that we are compelled to expend energy and incur other costs in the service of environmental modification. It therefore follows that the energy and other costs of achieving desired local environmental con-

[b]In the New York City Public Schools study, specific statistical comparisons were made in order to determine the effect of higher energy consumption on student performance. Standard reading/comprehension scores achieved by students in schools undergoing general modernization (including increase in lighting levels from 5–7 footcandles to 60 footcandles) were compared, both before and after the alterations, with those achieved by students in unaltered schools. The results were that the renovations had a short-term positive effect on student performance, but that disappeared by the end of a two-year period.

ditions can be minimized if we accept the existing natural local environment as a baseline to which we will add or subtract only to the extent strictly necessary to achieve the required conditions.

The basic architectural philosophy that derives from this energy conservative approach is that maximum harmony should be sought between the building and its environment. It should be understood that the harmony referred to is strictly physical and not necessarily aesthetic. Whether it follows the architectural aesthetics of Frank Lloyd Wright or Le Corbusier, a building may still be in *physical* harmony with its environment.

It might seem as though this point is so obvious that it barely needs stating. But that is not the case. The combination of continual improvements in the technology of environmental modification and cheap, abundant energy apparently seduced much of the architectural profession. Rather than adapting buildings to their environment, it became standard practice to use mechanical means to overpower any aspect of the environment that was in conflict with the architect's concept. For example, solar gain through wide expanses of glass was nullified by additional air conditioning, and heat losses through the same glass, by additional heating.

It is not at all difficult to understand this seduction. Improved capabilities for creation of artificial environments set the architect free of many of the environmental constraints that had previously made life difficult. The natural environment fluctuates, often unpredictably. Temperature, humidity, light, and wind patterns are notoriously changeable. It was therefore very inviting to seal the building against the vagaries of nature and provide a constant, controllable artificial environment. If natural winds were sometimes too strong and sometimes too weak, artificial winds were constant and congenial. If the sun had an annoying habit of blazing too brightly and then being precipitously darkened by cloud cover, artificial lights provided the desired predictability and uniformity. Not only windowless rooms but completely indoor streets for shopping districts were possible, immune to inclement weather. Artistic fancies were also indulged, and office buildings flooded with unneeded lights that shone through acres of glass battled for visual dominance of the night sky.

The sealed building became the ideal. Design of building service systems could be neatly "optimized" under the conditions of predictability that such a building offered. Computer simulations of building performance, which could not cope with the variabilities introduced by something as simple as operable windows,[c] functioned well within the sealed building context. The lure of computer simulation in allowing low cost, no risk testing of the performance impli-

[c]Richard Stein referred to this problem in his remarks to the energy seminar at Columbia: "Suppose one were to open a window to permit this exchange of inside and outside air to take place as one might have before we had sealed buildings. . . . Well, the computer can't compute it. We've checked with every program to see whether there is any way to simulate the performance of operable windows in a building and there isn't. The variability, the unpredictable quality of it is beyond the capacities of even very complex computer programs to simulate."

cations of modifications in building design was apparently extremely powerful. In the words of architect Stein, speaking primarily within the context of commercial public buildings,

> The result is that all of the recommendations for the energy use and environmental performance of buildings are always determined on the basis of the performance of a sealed building. It is accepted that this is a given.[6]

But the possibilities for computer optimization of building service systems for energy use were illusory. Because they optimized only within the context of the inherently inefficient sealed building, they, in essence, found the best of a set of bad solutions. Although buildings designed in such a manner may use energy very efficiently when energy inputs are strictly needed, because they are committed to using energy at all times, their overall energy performance tends to be quite poor. For example, the so-called "windowless school" in the Harlem area of Manhattan, a school with carefully designed and monitored building systems and thermostats in each space, uses two and one-half times the energy per square foot of the average of all other New York City public schools.[7]

The thrust toward ever greater stability of the internal environment favored increasing monitoring and control of building conditions by centralized machines. Machines were predictable, people were not. It was therefore only logical that control be shifted increasingly out of the hands of the individual building occupants. Not only couldn't windows be opened and closed as desired, but lighting heating, ventilating, and air conditioning systems were becoming less and less controllable by those for whom those systems were providing services. As elsewhere in modern society, this quest for predictability and reliability produced a terrible logic under which people were forced to adapt to the rhythm and characteristics of machines, rather than machines being designed to adapt to the rhythm and characteristics of people.

The ultimate result of this increasing disregard for the natural environment was a veritable orgy of energy use. The energy costs involved in sealing the building against some environmental effects and overpowering others were, and are enormous. Reducing the control of individual occupants over building systems with its inevitable correlate, "worst case" planning, led to massive wastage of energy. If lighting systems, for example, are not to be easily controllable by individual occupants, they must be designed to provide to the entire area sufficient light for the most visually demanding task that is ever performed in any part of an area. In addition, the light required for this "worst case" must be provided continuously and cannot allow credit for favorable natural lighting conditions that do not always occur. This type of design is a prescription for maximizing the needless consumption of energy. It is simply no longer tolerable.

By replacing the architectural philosophy of seeking to isolate and overpower the external environment with one of designing for maximal physical harmony between building and environment, large reductions in energy demand are possible with virtually no sacrifice in building performance. Some estimates have placed the potential energy savings at 50 percent,[d] which is likely to be conservative.[8]

There are, of course, greater constraints on the architect's artistic freedom implied by the philosophy of harmonious design. Yet one could argue that the greater the rigidities of structure within which a creative process operates, the greater the artistic achievement involved in creating within those constraints. For example, is not a poem like Robert Frost's "Stopping by Woods on a Snowy Evening" that much more impressive because it creates a gentle and thoughtful mood within the constraints of a very rigid meter and rhyme scheme? The challenge is greater, but when one reflects on the fact that virtually all styles of architecture that came into being prior to the last few decades could not have been created with a total disregard of the natural environment, it is clear the possibilities are considerable.

The Legacy of Primitive Architecture[9]

Before the advent of modern environmental modification machinery, architectural design had to adapt to the natural environment. The forms of architecture that developed over the centuries, particularly those outside of the spectacular traditions of houses of worship and public palaces, underwent an evolution that incorporated features designed to make use of the beneficial aspects of the natural environment while protecting against those that were hostile. Since these designs did not (because they could not) rely on large artificial energy inputs to achieve the desired effects, they may well supply us with a source of insight into techniques for achieving the sort of harmonious design we have been suggesting as an energy conservation strategy. Those forms which developed in harsher climatic conditions are likely to be of greater interest, simply because they had to cope with more severe environmental tests.

These architectural designs may be divided into two broad classes, *primitive architecture*[e] and *folk architecture.* The former term refers to the buildings of preliterate, pre-Iron Age handicraft societies, while the latter refers to all other buildings outside of the grand traditions of public and religious buildings. For example, Eskimo snow igloos, the tents of nomadic tribes, and the grass huts of

---

[d]Since the rate of energy-use in buildings has been doubling about every 10 years, this 50 percent energy differential is roughly equivalent to the differential energy consumption per square foot between a building designed in 1960 and one designed in 1970.

[e]The word primitive is used here without any implied value judgment. It merely refers to the earliness of the origins of such architecture, along with its relative simplicity.

sections of Africa are all classified as primitive architecture. On the other hand, log cabins, the houses of ancient Rome, and the courtyard houses of the dry tropics are considered folk architecture.

There is such richness and diversity of design in both primitive and folk architecture that it is absolutely impossible to discuss either in any real detail here. Therefore this section will focus on just a few specific examples, mainly of primitive architecture, in order to illustrate the high level of performance achieved by such buildings with little or no energy inputs beyond those provided by the natural climate.

The Snow Igloo

The extreme cold and wind of the arctic winter provides an extraordinarily harsh environmental test of building performance, but the Eskimo snow igloo meets this test extremely well. The snow igloo is a hemispherical dome, constructed from blocks of snow on the order of one and one-half feet thick, three feet long, and one-half foot high.[f] The shell thus constructed has considerable structural strength.

The shape of the dome has several advantages. It offers the minimum obstruction to winter winds, while at the same time having the lowest possible ratio of surface area to enclosed volume, thus minimizing the heat transfer area. In addition, a hemisphere is the volume most efficiently heated by a point radiant heat source. Since, aside from body heat, an animal oil lamp is generally the only heat source available, this shape is ideal.

The low thermal mass (heat capacity) of dry snow makes it an excellent wall material for a shelter exposed to continuous, intense cold. The fact that the wall does not absorb large quantities of heat means that heat produced by sources inside the igloo heats the air and internal objects without first having to fill up the heat reservoir of a high thermal mass wall. Thus the amount of energy input required to provide thermal comfort is held to a minimum.

The thermal insulative properties of the igloo are also good. The heat generated by internal heat sources in combination with the cold of the wall produces a melting and freezing phenomenon which results in the formation of a thin shell of ice on the inner wall. This ice film act as both a sealant and a radiant heat reflector. Radiant and conductive heat loss to the floor and walls is often further retarded by covering them with skins and furs. The net result of all this is a building in which, by virtue of the energy supplied by only a few oil lamps and human

[f]Typically, the snow blocks are laid in a circle, and the builder, working from inside the circle, lays row after row of blocks slanting inwards until the dome is completed. The only tool strictly required is a knife. See William A. Burns, *A World Full of Homes* (New York: McGraw-Hill, 1953).

bodies, the internal air temperature at the ceiling may be held as high as 65° F above external air temperatures.[g]

The snow igloo thus rates high in both structural and thermal design. However, there are some drawbacks. Though the ventilation is sufficient for breathing and sustaining a lamp flame, there is a considerable build up of odor. There is also a problem of melting when the outside temperature rises near the freezing point, say within 10°F. But, both because the igloo is often desired only as a temporary shelter by nomadic Eskimos and because it is so readily constructed, its impermanence is not a real disadvantage to those who use it. It is also interesting to note that there is absolutely no environmental pollution associated with the practice of abandoning these naturally degradable structures.

The Architecture of the Low Latitude Deserts

In the deserts of the lower latitudes the climate is characterized by dryness coupled with very wide diurnal and sometimes, seasonal temperature variations. The primitive and folk architectural response (visibly in Egypt, Mesopotamia, Mexico, and the Southwest United States) has been characterized by massive walls, narrow windows, and roof terraces. Though there is considerable variation in the structures, these features serve as a common bond between structures used by cultures as diverse as the Pueblo Indians of the United States Southwest, the rural villagers of upper Egypt, and urban dwellers of sections of Morocco.[10]

The massiveness of the walls, combined with the fact that they are invariably made of high heat capacity materials (e.g., mud, clay, molded earth, adobe brick) gives them very considerable thermal mass. They are thus able to absorb much of the intense solar radiation of the daylight hours and reradiate it gradually at night. As a result, the roof and walls act as a thermal buffer, protecting the interior from both the excessive daytime heat and the chill of the night air.

The few windows are kept small to avoid high solar heat gain, as are the doors. Since cutting down on the entry of solar radiation in this way also reduces the penetration of light, interior walls are light colored in order to reflect rather than absorb the light that is admitted. In this way, interior areas are kept much brighter than would otherwise be the case. In fact, the light coloring of interior and exterior walls improves their performance as reflectors of radiant heat as well.

The building is oriented relative to the sun and the prevailing winds so as to allow maximum ventilation with minimum solar gain. In more sophisticated folk

---

[g]However, since it is not unusual for the outside air temperature to be as low as -30°F, even this large a differential will not result in an internal air temperature much above the freezing point of water. See J.M. Fitch and D.P. Branch, "Primitive Architecture and Climate," *Scientific American* (December, 1960), p. 138

architectural designs, ventilation shafts that catch the prevailing wind and channel it to interior rooms may be included.[11] In primitive architectural designs, the arrangement of windows performs a similar function.

The net effect of the combination of these thermal design factors is a very considerable smoothing of the diurnal temperature fluctuation. A primitive adobe house, for example, transforms a typical 65–105° F external daily temperature range into a 75–85°F internal daily range of temperature without any nonclimatic energy input.[12]

The nature of the climate in the hot dry tropics is such that exterior living is often as important as interior living. There are two particular aspects of these structures that facilitate outdoor living: roof terraces and courtyards. The roof terraces make excellent sleeping quarters in areas where nocturnal temperatures are not too low. Some warming effect is produced by the radiation of heat from the roof to the night sky. Interior courtyards, a sophistication of folk architecture, are shaded during the day but open to the sky at night. Typically, in the courtyard tradition of urban areas, such as Baghdad, Iraq, living rooms open to the courtyard and not the street, and the courtyard is an important part of the structure's space.[13]

In arid regions that have seasonal rains, flat "sleeping roofs" are not feasible. Instead, water shedding, sloping roof surfaces are necessary. In Nigeria, for example, double-shelled "beehive" huts are constructed. The inner mud dome is separated from the outer thatch dome by an air space. The water shedding outer dome protects the inner mud dome during the rainy season, the air space serves as thermal insulation, and the high thermal mass inner dome performs the function of smoothing the diurnal temperature fluctuation.[14]

The Architecture of the Tropics

The earth's tropical zones are characterized by intense solar radiation and little temperature variation, either seasonal or diurnal. The lower latitude tropics, however, have heavy rainfall and high humidity all year around, while the higher latitude zones typically have one season of high rainfall and one that is quite dry. Thus the structures of the tropics must, in all cases, deal with high heat, but must function under very different rainfall and humidity conditions.

A common feature of the primitive architecture of these tropical zones are minimal walls typically constructed by tying or weaving of vegetable fibers around frames made of bamboo, saplings, etc. The roofs, made of the same materials in the same manner, are steeply sloped and often extend well beyond the inner living space, dominating the structure. In addition, floors are sometimes raised significantly above the ground level on stilts.

Consider the engineering aspects of such structural designs. The vegetable materials of which the buildings are constructed have fairly low thermal mass, reducing the build-up of heat. Since the diurnal temperature fluctuations are small, any such heat accumulation could not be readily dissipated to a cool night sky. For the same reason, the walls need not perform any temperature smoothing function. Thus low thermal mass is highly desirable.

The minimal physical mass of the walls further reduces their heat capacity while increasing interior ventilation. In some cases, walls are completely eliminated to heighten both these effects. Where significant walls are present, for example, in the village houses of Luzon in the Phillipines, windows are placed on all sides of the structure. In such houses, windward openings can be closed during rainfall with leeward windows left open to provide ventilation.[15]

The large umbrella-like roof shades the interior, protecting it from the intense sun. Both the construction and size of the roof perform important thermal functions. The woven fiber mat construction of the roof renders it essentially opaque to solar radiation, therefore heightening its shading effect. The roof's extensiveness guarantees that the air and ground space immediately outside the inner living space will at least lie in the penumbra of the shadow it casts most of the time, thus reducing the convective and radiative heat gain to the interior. Besides protecting the interior from the oblique rays of the sun, this overhanging roof also affords some shelter from the wind-driven rain.

Placing the floor on stilts performs several important environmental comfort functions. It improves ventilation by raising the house above low level ground obstructions so that it may catch the wind more efficiently. It also reduces the thermal mass of the floor by replacing a high heat capacity earth floor with a low heat capacity fiber floor. In addition, the free air flow under the floor performs a further cooling function, while the separation of the floor and the ground reduces ground-level water problems.

In the biseasonal areas of the tropics, there are some very interesting modifications of the basic humid-tropic primitive house design. For example, certain South African tribes build a light wooden frame hut covered with woven mats. During the dry season the weave of these mats contracts, allowing air movement through the walls. When wet weather comes these fibers absorb moisture and expand, tightening the weave encough to essentially prevent the rain from entering. In some cases these mats can also be moved from wall to wall in accordance with the direction of the wind.[16]

The Lessons of Primitive Architecture

From the viewpoint of the standards of human comfort and internal environmental performance to which we have become accustomed, all primitive structures leave something to be desired. In some cases it is the buildup of odors due to poor ventilation, in others the impermanence of the structure, in still others the unacceptably low level of internal illumination. They are far from ideal buildings. But they do have some important things to say from the viewpoint of energy conservation.

Probably the single most important lesson to be learned from primitive architecture is simply that it is possible to design buildings capable of achieving relatively impressive levels of internal environmental performance without requiring substantial inputs of nonclimatic energy. In a sense, these structures constitute a sort of empirical "existence proof." However, they do much more than that, because they also represent case studies in the application of some of the most critical principles of architectural energy conservation.

No one in his or her right mind would advocate the widespread exact replication of primitive forms as a real alternative for the architectural needs of modern societies. But there is absolutely no reason why the basic lessons of design-to-climate, like attention to thermal mass, building orientation, window arrangement, building shape, etc., cannot be applied to modern structures. There is a wide range of construction materials and techniques available to contemporary architects and building engineers that are in many respects far superior to anything available to builders of primitive structures. Therefore the design of modern structures that equal or surpass the energy efficiency of those of primitive architecture should present no insuperable problems. The development of those designs represents a real challenge to the ingenuity of modern building designers, but it is a challenge that we have every reason to believe can be met.

Strategies for Energy Conservative Building Design

Study the Macro–and Microclimate

A detailed knowledge of the climate in which a building will function is an absolute prerequisite to the design of an environmentally harmonious building. The normal macroclimatic data available from weather stations, while important and useful, is simply not detailed enough to allow for sufficient tuning of the building's design. Area-wide data does not pick up all of the microclimatic variations that invariably occur because of local geographic features and buildings within the immediate vicinity of the building site. As Hay has written, within the context of residential design, "Detailed data on temperature, dew point, precipitation and cloud cover, and solar radiation are prerequisites for the selection of house design and materials and for consideration of heating and cooling."[17] Of course, the more specifically these data apply to the particular site upon which the building is to be constructed, the more completely the building can be harmonized to the environment.

As a matter of perspective, it is important to understand that a great deal can be done in the way of environmental architecture with only macroclimatic data. However, since the building is stationary, microclimatic data is more relevant to the actual conditions under which the building will operate. It therefore allows a greater saving of energy. Of course, fine tuning can be overdone. The fact that micro-and even macroclimates change over the typical lifespan of most structures indicates that designs that are ultra sensitive to minor changes are to be avoided.

Maximize Net Positive Energy Flow

Following the terminological conventions of Stein,[18] any flow of energy through a building skin due to natural forces that would otherwise have to be artificially supplied is considered positive. Similarly, any such natural flow of energy that must be countered by additional artificial energy expenditure is considered negative. For example, a natural removal of excess heat from a building in summer constitutes a positive energy flow, while a natural loss of needed heat from a building in the winter constitutes a negative energy flow. In short, an energy flow that is beneficial, from an energy conservation viewpoint, is called *positive,* and one that requires additional "makeup" energy expenditure is *negative.* These terms positive and negative do *not* refer to the direction of flow per se.

It should be clear that greater energy savings are achieved the higher the net positive energy flow. Thus energy conservation calls for maximizing the net positive energy flow subject to all the environmental and technical constraints. However, this has clearly not been standard architectural practice, as architects Richard and Carl Stein have pointed out:

> Over the past thirty years, devices and systems which modify the internal environment have tended to be mechanical, requiring the input of energy for their operation. In designing these systems, the negative energy flow through the building's skin is calculated, since it is this that the mechanical system must offset; however, the positive flow is usually ignored. . . . This attitude has led to a general point of view which considers building skin (or perimeter) a necessary evil which should be minimized. . . every attempt is made to neutralize the energy flow between the inside and outside environments thereby creating a relatively predictable interior situation and to provide the environmental factors by mechanical means with reductions in fuel usage resulting primarily from the optimization of systems and the minimization of negative energy flow through the skin. This approach results in buildings which are totally dependent on their mechanical systems.[19]

Buildings designed in this way are committed to continuous and often redundant energy use. They are therefore inherently inefficient users of energy, no matter how efficiently their energy-using systems operate when considered in isolation. But if, instead, buildings are designed to maximize net positive energy flow, their energy-using systems are called into play only when that natural energy flow is insufficient to meet environmental requirements. The building's systems become supplemental rather than redundant. Thus energy is consumed only to the extent strictly necessary.

Avoid Uniform Worst Case Design

Both the environmental and structural demands on virtually all buildings are subject to considerable variation. The structural stresses typically vary substantially, even along the length of a single beam. Lighting, heating, cooling, and ventilation requirements vary from place to place within a building at any given point in time, as well as diurnally and seasonally. Within this variation, the most demanding extreme is called the *worst case.* The ventilation requirements of bathrooms and kitchens during periods of peak use, the lighting requirements for visual tasks such as drafting, the arrangement of internal equipment and personnel that puts the maximum load on the most heavily loaded portion of a beam—all these are examples of worst cases.

Obviously, it is wise to design the structural and environmental systems to be capable of meeting the requirements of the worst case without excessive strain or damage. But since these worst cases tend to be quited limited both in location and in duration, there is no reason why the design should call for uniform provision of worst case requirements over elongated periods of time and extended spacial areas. However, this is precisely what has been done.

It is, of course, always easier to specify a uniform level of lighting, ventilation, heating, load-bearing capacity, etc. than it is to account for all the temporal and spacial variations that occur in actual requirements. And since worst cases do occur, albeit only in limited areas and/or for a short periods, the "safest" uniform level of services to specify is that required to meet these most demanding situations. In fact, it is even safer (and frequently legally necessary) to specify levels several times higher than these—and this is very often done.

The difficulty is that such designs impose tremendous energy penalties. In construction of a building, the pileup of worst case structural safety factors results in an excessive use of materials. Since large amounts of energy are used in the manufacture (and to a lesser extent in the transportation and placement of such materials), excessive materials mean excessive energy-use. Likewise, worst case design of the building's environmental systems leads to considerable overprovision of lighting, heating, air conditioning, and ventilation and, accordingly, excessive energy-use in the operation of the building. The vast overdesign of environmental systems is undoubtedly the greater problem of the two, although both are quite serious, because the unnecessary consumption of energy it represents persists over the entire life span of the building.

By way of illustrating the extent to which structural safety factors based ultimately on worst case considerations can pyramid, consider the following calculation based on Stein's example of the design of a single school classroom.[20] According to the National Building Code,[21] a 750 square foot classroom designed for 30 students and a teacher, must be capable of withstanding a load of 40 pounds per square foot, *on every square foot.* This load calculation is in addition to that required by the weight of the structure. It therefore provides for a teacher/ student/ equipment load of 30,000 pounds. If one were to make the extremely conservative assumption that the average body weight of the thirty-one occupants was 175 pounds, the total body weight load would be 5,425 pounds. Assuming, again conservatively, that each desk and chair combination averaged 100 pounds, another 3,100 pounds is added. Allowing about another 1,500 pounds for other classroom equipment, books and additional people, yields a total load of approximately 10,000 pounds, or one-third of the computed 30,000 pound requirement. But since design computations assume a value of strength of steel only about half of its actual strength, and a value of strength of concrete only about one-third of its actual strength,[22] the safety factor is not merely three, as it would seem from the load calculations, but rather six for steel and nine for concrete! In fact, Stein points out that the safety factors are even higher than this because,

> There are additional safety factor margins provided by selection of dimensions above these computed, by the choice of concrete above design strength, by the complex way in which a structure reacts to loading, and by the fact that concrete gains strength for years after hardening.[23]

By reducing these obviously excessive material requirements, the weight of the structure itself would be reduced, thus allowing a further reduction in materials. Stein estimates that elimination of half the concrete now used in building construction is perfectly reasonable and safe, and that such a reduction would save 20 billion kilowatt-hours of energy per year in cement production alone, an amount roughly equal to the electric power consumption of nearly 3 million families.[24] To this would be added energy savings in steel manufacture and transportation of steel and concrete.

These savings could only be achieved at the expense of increased expenditure in time and effort for much more careful design and construction. In the design stage, it would be necessary to take account of the variations in structural loading one could reasonably expect to occur and to provide safety factors in some sensible relation to the actual loads each portion of the structure will encounter. During construction, more labor would be required to see that more care is taken in building of formwork and in preparation and placement of the various structural components. However, it is worth keeping in mind that energy is not all that will be saved by these procedures. The pollution of the environment and the depletion of other resources associated with the production and transportation of construction materials will also be reduced.

It is always difficult to advocate the reduction of safety factors. But is it really safer, in any meaningful sense of the word, to have a structural safety factor of 900 percent rather than say 500 percent? And if so, why stop at 900 percent? Why not 1,500 percent or 2,000 percent? Obviously there are both technical and economic issues here. The technical issue revolves around the determination of the worst case loads that each different part of the structure might encounter with nontrivial probability and the degree to which excess structural capacity beyond that required to meet these worst case loads contributes to the structural performance of the building under both normal and extreme stresses. The economic issue, on the other hand, is preeminently a matter of tradeoffs between the added materials and energy expense of achieving acceptable levels of technical safety by using uniform worst case design and pyramiding safety factors, and the added labor expense of achieving technical safety by more careful design and construction.

In the era of cheap energy, acceptance of waste, and environmental disregard, the economics of the situation dictated the excessive materials approach. But the essential point is that this approach is patently not dictated by the fundamental technical situation. It is easily within the bounds of present technical knowledge to achieve the same levels of structural performance in buildings with enormous savings in materials and consequently in energy.

Since the environmental side of the worst case design problem is dealt with in some detail in the succeeding chapters, it will not be discussed further here. However, it is interesting to note that in contrast to the structural design situation, avoidance of relatively commonplace worst case environmental design would, in many instances, have resulted in considerably monetary and energy savings over the building's life cycle, *even at pre–1973 energy prices.* Thus we must look elsewhere for an explanation of why such monetary gains were not exploited.

Simpler Is Better

It is a basic tenet of engineering design that the simplest design that will satisfy performance requirements is the best. Complexity is accepted only to the degree that it is unavoidable. The basic technical reason for this is that complexity breeds unreliability. The multiplication of nonredundant components in a system tends to rapidly reduce overall system reliability even in the face of substantial increases in the reliability of individual components.

As a simple illustration, suppose a system (e.g., machine, building, etc.) can be designed either of two ways. The simple design consists of two components, each of which is 80 percent reliable (i.e., has a 20 percent probability of failure); the complex design consists of five components, each one of which is 90 percent reliable. Assuming that all of the components in either design must function properly for the system to meet performance requirements, and further assuming that weakness in one component does not alter the probability of failure of other components, the probability of failure of the complex system would be 41 percent, while the probability of failure of the simple system would be only 36 percent. And this despite the fact that every component of the complex system is considerably more reliable than any component in the simple system. Although the problem would be considerably more involved if weakness in one component did increase the probability of failure of the other components, as is typical in real world cases; this would tend to increase rather than reduce the advantage of the simpler design.

Simpler systems tend to be less sensitive to minor modifications of the environment in which they operate. They therefore tend to perform better under less constant environmental conditions. Simpler systems are also generally easier to repair.

In addition to the technical advantages of simplicity, there are also some important economic advantages. Simply designed systems are typically cheaper to manufacture and thus cost far less. Although complex systems can usually be rendered more reliable by designing in redundancies (i.e., back-up systems), thus undoing some of the reliability losses of complexity, these redundancies add considerably to the cost of the system. Maintenance as well as repair costs are often higher for more complex systems. Thus, as a general rule, there are excellent economic and technical reasons for preferring simpler to more complex designs, among those which satisfy essential system performance requirements.

This general bias toward simplicity is as valid for buildings and their internal systems as it is for any other engineering systems.[h] The difficulties of complexity as they relate specifically to buildings, are illustrated by one of the chief findings of the detailed Stein study of the thousand building New York City Public School System,

> In general, the more complicated sophisticated systems functioned far less effectively than the simpler more common ones. Buildings that relied entirely on mechanical delivery systems for their interior conditions appeared to have the greatest number of problems with control systems, to have the widest temperature variations, and to consume the most fuel.[25]

It must be emphasized that the goal of energy conservation is not served by a blind preference for simplicity. There are many important instances in which the installation of some additional control devices, for example, will have handsome energy conservation payoffs, so that the added complexity is well worthwhile. Likewise, the careful attention to specific external environmental

[h] Simplicity in engineering design is completely different from simplicity in artistic design. Both artistic complexity and simplicity, whether in the appearance of a building or a machine are compatible with simplicity of engineering design.

conditions and to the variation in actual structural and internal environmental demands advocated in previous sections adds complexity to the design and construction process. But this is thoroughly justified by energy considerations. A bias toward simplicity does *not* imply simplicity at all costs. Rather it does imply viewing every increase in complexity with a jaundiced eye and requiring that each additional complication be rejected unless its incremental benefits are substantial.

Increase Occupant Control of Environmental Systems

The human body contains highly developed environmental sensors that are capable of detecting alterations in temperature, humidity, air movement, light, odor, noise, etc. with considerable sensitivity. However, human homeostatic systems, in combination with involuntary sensory adjustment mechanisms, operate sufficiently well to allow adaptation to minor changes of this type without conscious awareness or voluntary action. Thus apparent or conscious sensitivity to variations in environmental conditions is substantially lower than ultimate physiological sensitivity. For example, the normal human eye is capable of adapting automatically to maintain visual comfort under variations in light many orders of magnitude larger than the minimum flash it is capable of perceiving under laboratory conditions.[i]

There are essentially three ranges of environmental alteration definable in terms of human sensitivity and response. Within the first range, automatic adaptation to environmental changes takes place without any conscious awareness that anything has changed. Beyond that range is a second range within which the person perceives that a change has taken place but rapidly adapts to the new conditions involuntarily and reestablished comfort. The third range includes changes so large that they are not only perceived but require conscious adaptive action.

To continue with the visual example, consider the case of reading a book by natural light on a cloudy summer's day. There will be continual variations in the intensity of light falling on the page as a result of the continuously changing solar and atmospheric conditions. Most will be so minor that the lighting level will appear unchanged (first range). Not infrequently the page will appear to get brighter or darker, say with variations in the thickness of the cloud cover, but after brief discomfort the eye will adapt to the new lighting condition and the reader will experience no discomfort in continuing with the reading (second

---

[i]The minimum perceptible flash of light under carefully controlled experimental conditions has been calculated as 54–148 photons (quanta of light energy). As little as a single photon may be enough to initiate a photochemical reaction in one rod of the eye, though apparently 5–14 rods must be stimulated for the event to be perceived as a flash of light. See Robert S. Woodworth and Harold Schlosberg. Experimental Psychology, rev. ed. (New York: Holt. Reinhart and Winston. 1965). p. 377.

range). However, periodically the sunlight will suddenly burst brightly through a hole in the cloud cover and illuminate the page so brightly that the reader must either put on sunglasses, provide shade for the page, or else temporarily cease reading in order to reestablish visual comfort (third range).

Operationally, this implies that within the first two ranges there is no need to monitor and adjust environmental conditions to maintain artificially constant levels of temperature, humidity, lighting, etc., using energy to compensate for natural variation. Unless natural variations within the second range are so frequent and so wide as to stress the adaptive mechanisms and produce undue fatigue, this compensation constitutes a wastage of energy. Furthermore, some natural variation tends to be pleasant, physiologically and aesthetically.

Since people are fully capable of detecting and initiating conscious corrective action for variations in the third range, they should be given the ability to do so. Excessive automation of building environmental systems is at once unnecessary, unpleasant, and highly wasteful of energy. The more flexibility people have in the adjustment of their environment, the more they are capable of avoiding energy wastage by eliminating redundant or excessive mechanical provision of environmental services. Beyond this, different individuals have different tolerances for environmental variations of various sorts, as well as different comfort levels. To some extent this is the result of cultural conditioning, but to some extent it is also physiologically and psychologically determined. The ability to adapt one's own environment to one's own needs provides the potential for maximal comfort with minimal energy consumption. It is also psychologically beneficial, providing some counter to the feeling of being dominated by forces outside one's own control, of being a cog in a machine.

As an illustration of how not to design a building's environmental services, I would like to offer the example of my office at Columbia University. The office, located in a relatively modern building (completed in 1961), is roughly a rectangular solid in configuration measuring approximately 20 feet by 9.5 feet by 11.5 feet. It is equipped with two parallel banks of eight 4-foot, 40-watt fluorescent lights. There is only one light switch controlling the lights, so that either all 16 bulbs are on or they are all off. There is one window measuring 5.5 feet by 4.5 feet at the short end of the rectangular area, with an operable drape. The window is not actually sealed, but it is set back about 2 feet from the wall, in such a manner as to render it difficult to operate. The frame is also constructed so that it is not easily operable. To date, I have only been able to operate the window by jumping up onto the window ledge. The window is clearly not intended for casual operation.

There is a forced-air system that continuously blows air from early in the morning until 10 P.M., whether or not the room is occupied. The system can neither be adjusted nor shut off by any readily accessible control inside the room. The air is heated or cooled to a temperature that also cannot be adjusted by the room's occupants. The excessive height of the room increases the volume of space to be heated/cooled and ventilated by more than 40 percent. The room is occupied only a fraction of the week during the academic year and even less frequently during the summer. The total shutdown of the forced-air system after 10 P.M., combined with the functional inoperability of the window, makes working conditions in the room quite uncomfortable beyond that hour.

Aside from control over the electric lights, the ability to open or close the drape, and open or close the door, there is no effective control over the provision of any of the environmental services in the room. The result is a considerable wastage of energy in frequently unnecessary or excessive heating, cooling, ventilation, and lighting of the room. A few simple switches and operable control devices (and a lower ceiling) would drastically reduce this energy wastage and improve the quality of provision of environmental services. The additional cost of providing these devices during the original design and construction of the building would have been small, the energy payoffs, very large.

Although people clearly have the capacity to monitor and adjust environmental conditions so as to conserve energy, there are a number of reasons why they should not be given complete control of environmental services. For one, because other activities require their attention, people are distracted from the energy monitoring function. After all, the provision of environmental requirements should establish background, a context within which the normal business of living is carried out. If it requires too much attention, it becomes burdensome. Secondly, a certain basic level of environmental services may, for reasons of safety, be best provided independent of the control of individual occupants. Lighting in stairwells and hallways, and to some extent minimal ventilation, fall into this category. A third reason is the tendency of people to be careless about turning off unneeded systems (e.g., lights), particularly where they do not directly bear the costs of the consequent energy usage. Finally, in some cases occupants may either not be physically capable of making normal environmental adjustments (e.g., some patients in hospitals) or may not be responsible enough to be allowed free operation of such controls (e.g., some students in schools).

Considering both the advantages and disadvantages of allowing individual occupant control over the local provision of building environmental services, the general rules for establishing an optimal energy conservative mix of manual and automatic controls would appear to be as follows:

1. Thermostats and other control devices should be provided for each individual space. These devices should be adjustable by the occupants within a broad range.
2. All environmental monitoring control devices (e.g., thermostats) should be capable of maintaining levels of services with variations primarily in the first and perhaps occasionally in the second range without further attention once they have been activitated and set by occupants.
3. Occupants should be able to totally shut off, or at least reduce to minimum

safe level (where such a level exists), all environmental systems in their local space.

4. All environmental services required at minimal levels, which for safety purposes should not be casually controllable, should be provided automatically. Monitoring devices that automatically activate or deactivate such systems so as to avoid redundancy with naturally provided conditions should be considered. At the very least, timers should be employed for this purpose where more sophisticated systems are judged impractical.
5. Windows should be operable, where safety allows, and should be provided with operable drapes, shades, etc.
6. Automatic shut off (or reduction to minimum safe level) of all environmental services should be provided by timing devices to correct for occupant carelessness and therefore avoid before and after hours energy wastage. However, temporary manual overrides to all such devices should be provided. As an alternative, security guards on patrol and/or janitorial personnel could manually perform the same function.

As compared with current widespread practice in design of nonresidential structures, these guidelines constitute a call for considerably increase in occupant control. The net result of following these suggestions should be both an increase in the quality of building services through a greater tailoring of these services to actual needs and an enormous saving in energy.

Adjust Building Orientation for Energy Conservation

Both external and internal building orientation are relevant to the energy consumption of buildings. *External orientation* refers to the orientation of the building itself on the site as well as the arrangement of features of the building skin (e.g., windows). *Internal orientation* refers to the arrangement of functional spaces within the building.

External orientation relative to winds and sun can be critical to providing natural ventilation and lighting, and to minimizing or maximizing solar gain, whichever is desired.[j] To allow the placement of a building on its site to be determined solely by the position of streets is not only to perpetuate the mistakes of the past but also to foreclose some important energy conserving design-to-climate options for the life of the building. Similarly, placement of windows and doors is far too important to be determined strictly on the basis of aesthetics or architectural whimsy. For example, the simple fact of placing a loading dock on the

[j] If environmental conditions are appropriate, one might even achieve such pleasant effects as natural clearing of the bulk of snow from driveways and entrances by orienting the building properly relative to prevailing winter winds.

leeward rather than the windward side of a building can save considerable loss of heated or cooled air and thus energy if the dock is large and frequently used.

The arrangement of functional spaces within the building also has important energy consumption implications. Perimetric spaces should be allocated to those activities that can most benefit from closeness to the building skin (e.g., classrooms, offices). Those activities for which the presence of windows is either irrelevant or detrimental (e.g., elevators, storerooms) should be assigned interior spaces. Placement of activities that generate heat near the building perimeter facilitates the dissipation of that heat, interior placement facilitates heat transmission to other internal areas.

Use Less Energy Intensive Building Materials

To the extent that building materials are substitutable, it would be highly desirable from a social view point if the choice of building materials were made with serious consideration given to energy implications. There are substantial differences in the energy content of various building materials—differences resulting from differing energy requirements of both manufacturing processes and transportation.

In order to determine which of two substitutable building materials has the least energy content, it is important to consider not only the energy required to manufacture (and transport) a ton of each material but also to calculate how much of each material is required to perform the same building function. Clearly no energy savings will result by switching to a material that can be produced with half as much energy per ton if twice as many tons of that material are needed.

Summary and Conclusions

Architectural design is a key determinate of the energy performance of any building over its entire life cycle, strongly influencing the energy required for proper operation of building service systems as well as the amount of energy built into the structure itself. In order to minimize the consumption of energy in the provision of acceptable internal environmental conditions, the building should be designed in harmony with its external physical environment. Although this puts some constraints on architectural choice, it should not be excessively restrictive, since there is no reason why the physical harmony of building and environment should not be compatible with a very wide range of aesthetic architectural philosophies.

Primitive and folk architecture developed before the advent of modern environmental modification machinery had to be designed with close attention to climatic conditions. Many of these designs were (and are) remarkably effective, providing internal conditions far more congenial than external conditions, with little or no consumption of energy resources. There is much that energy conservationists can learn from the study of such structures—much that should be adaptable to modern architecture with the application of some ingenuity.

A prerequisite for the application of the design-to-climate approach to architecture is detailed knowledge of the macro-and microclimate at the building site. Once such knowledge is obtained, the designer should seek to maximize the net positive energy flow between the building and its environment. Uniform worst case design and its correlate, multiplication of safety factors, should be avoided, both with respect to structural and building systems design. Biasing the choice among alternative designs in the direction of simplicity should achieve energy savings as well as increased reliability. Increasing occupant control over environmental systems by limiting central control to a few minimal functions (mostly related to safety) should also provide substantial energy savings. Attention to orientation of the building on its site and to the arrangement of elements of the building skin, along with appropriate orientation of functions within the building, are other architectural considerations important to the conservation of energy. So too is the choice of less energy intensive building materials.

It is difficult to estimate with any degree of precision what the total net impact on energy consumption would be if the whole complex of strategies advocated here were conscientiously applied. But there is little doubt that it would be substantial. The estimate that it should be possible to achieve a 50 percent reduction in energy-use in buildings over the current levels seems reasonable, if not conservative.

3 Heating, Cooling, and Ventilation

More than one-quarter of the total energy consumed annually in the United States is used for space and water heating, cooling, and ventilation.[1] Much of this energy is wasted through a combination of excessive environmental design requirements, the necessity for overcoming unwanted heat loss/gain resulting from poor structural design, inappropriate selection and design of building systems, and inefficient operating practices. Part of the elimination of this wastage requires only relatively minor building modifications or more careful attention to energy management and may thus be achieved within the context of existing buildings. But part requires rather basic redesign and so is primarily applicable to new structures. Accordingly, though a very significant reduction in energy consumption for heating, cooling, and ventilation is both technically and economically feasible in the relatively short run, the full potential of the energy savings will be realized only in the somewhat longer run.

Preventing the Unwanted Loss or Gain of Heat

Insulation

Thermal insulation inhibits the passage of heat by virtue of its low thermal conductivity. It can therefore be used to hold heat where it is desired, avoiding the necessity of consuming additional energy resources to cope with the higher air conditioning or heating load implied by the dissipation of that heat. Thus insulation should be applied wherever contact between two systems at differing temperatures cannot be avoided and heat transfer between them is considered undesirable.

As a general rule, all pipes containing hot or cold substances should be well insulated except, of course, at any points where the transfer of heat is specifically desired. For example, pipes carrying hot water to be used for cleaning purposes should be insulated along their entire length, while those carrying hot water intended for space heating should be insulated everywhere but at the points where they traverse the spaces they are to heat. Likewise, boilers and furnaces should be thoroughly insulated.

Insulation applied between the walls of buildings can drastically reduce the heat gain in hot climates or heat loss in cool climates, without any obvious external or internal aesthetic effects. Roof insulation has similar capabilities. Insulation applied to the ceilings and floors prevents the heating or cooling of the service areas between stories as well as the transfer of heat between functional areas that may have very different thermal requirements. It has been estimated that the use of only 3 inches of thermal insulation in walls and floors and 6 inches in ceilings can reduce the heat loss/gain in residences by as much as 50 percent.[2]

There are both technical and economic aspects to the question of how much insulation is optimal. On the technical side, additional insulation tends to increase the outer area of the exposed surface. Since radiative heat loss tends to be larger the larger the final heat-losing surface, the benefits of added thermal insulation may be partially offset by this increase in surface area. This effect is particularly important with respect to small diameter pipe. Wrapping a ½-inch bore pipe in 1 inch of insulation quintuples its area. It is conceivable, particularly if the insulation is of poor quality, to reach a point beyond which additional insulation may actually increase heat transfer.

In general, the technical efficiency of insulation is subject to a kind of diminishing marginal return. For example, a 6-inch bore steel pipe carrying steam at 600°F (through still air at 70°F) will have a temperature of 155°F at the outer surface of a 1-inch thick layer of insulation. Successive ½-inch increments in the thickness of insulation will drop the surface temperature by 20°F, 13°F, 10°F, and 8°F, respectively.[3] In addition, the absolute magnitude of change in the difference between outer surface temperature and internal temperature produced by comparable successive increments of insulation is smaller, the closer the internal temperature is to the ambient temperature. That is, if a pipe carries materials only slightly warmer (or cooler) than the surrounding temperature, additional insulation will not be very effective. Similar effects obtain for surfaces other than pipes.

The technical situation has two main implications. First, *it is far better to insulate every part of a surface* (e.g., a pipe) *to some degree than to highly insulate part of it and leave other parts bare.* Second, if it is necessary to trade off, for reasons of space or economics, between insulation on one surface and insulation on another, the *insulation should be more heavily applied to those surfaces that separate areas with greater temperature differentials.* For example, pipes carrying high-temperature steam or water or pipes carrying cold refrigerants should be favored over pipes carrying warm or cool water.

The costs of insulation vary more or less directly with the amount of insulation installed. However, the pattern of this variation is very difficult to specify more precisely than this for the general case because, though the cost of the insulating material itself tends to vary linearly with the amount used, the variation in the labor and other costs is more complex, especially in the case of existing buildings. Even though generalization is difficult, it should not be particularly difficult to estimate the costs of insulation in any specific case.

The combination of monotonically increasing cost and diminishing marginal technical efficiency of added insulation implies that there is an economically optimal amount of insulation. There are a number of ways one could determine this optimum for any specific case. First, both the cost of each level of insulation and the resultant fuel savings should be computed. Then, using the internal rate of return criterion, for example, one would compute the rate of return implied for the additional investment required for each successive level of insulation and choose the largest amount of insulation that yields a rate of return at least as great as is obtainable on alternate investments.[a]

The only part of this calculation that poses real difficulty is the estimation of fuel savings. The problem here is not the estimation of quantity of fuel saved per year, rather it is the estimation of future fuel prices. Using present fuel prices will result in a serious underestimate of fuel-cost savings if energy prices continue to rise. Thus such a calculation will be biased in the direction of under-insulation.

It is important to note that the optimum discussed here is neither necessarily the same as the technical optimum nor the social optimum (whatever that may mean). It is purely the narrowly conceived economic optimum from the viewpoint of the individuals investing in the building and its systems.

Thermal insulation is at once one of the simplest, cheapest, and most effective means of energy conservation available. It is applicable to existing as well as to new structures and often provides an attractive financial investment opportunity. Insulation practices, especially in residential structures and industrial facilities, have been so poor in the past that very large energy savings should be derivable from applying even relatively low standards of insulation universally.

Fenestration

Windows represent one of the most critical areas for the transfer of heat, light, and air between the building and its environment. They afford a view of the external world that can greatly enhance the psychic benefits conveyed by the building to its occupants. Properly designed, windows operate as valves which can be used to control the flow of energy between the internal and external environments. As such they play a key role in attaining the goal of maximizing the positive (or minimizing the negative) net energy flow. On the other hand, improperly designed windows constitute "holes" in the walls, holes that can undo the effects of the most thorough and carefully applied structural thermal insulation.

---

[a] This will not in general be the highest of the rates of return computed. One should not stop, so to speak, at the point of maximum return, because as long as the rate of return derivable from additional expenditure is greater than that rate available elsewhere, the total return derived from all investments will be increased by investing still more funds in insulation.

Type of Glass. Standard, quarter-inch clear plate glass is essentially transparent to solar energy in the visible portion of the spectrum but relatively opaque to longer wave radiation. Via the previously discussed greenhouse effect, much of the solar energy transmitted through such glass is experienced as interior heat gain. In addition, glass of this type has a thermal conductivity five to ten times as great as that of reasonably well insulated walls.[4] The combination of these properties implies excessive interior heat gain in the summertime, both from incident sunlight and from conduction of exterior heat to the interior as a result of the air temperature differential. In the wintertime, the beneficial effects of high solar heat gain tend to be more than offset by the leakage of heat through the highly conductive glass. Thus ordinary plate glass has some serious drawbacks as a window material from an energy conservation standpoint.

Fortunately, there are several ways to ameliorate these difficulties. Probably the simplest is to double or triple the amount of glass in each window. The most effective and practical way to incorporate this additional glass into the design is to use two or three sheets of glass separated by layers of air. The air layer serves an insulating function without distorting or in other ways disturbing the visual effectiveness of the window. A second technique is to coat the glass with an appropriate material to enhance its insulative or reflective properties. There have been a number of reflective metallic coatings introduced in the late 1960s and early 1970s that offer the capability of blocking solar heat and light with a broad range of degrees of effectiveness. A third approach would be to search for materials other than glass which have the desired reflective, insulative, and visual properties. Various clear plastics have, for example, been used in place of glass for skylights. However, one must exercise great caution in the selection of alternate materials to avoid those which require inordinately high consumption of energy resources to manufacture.[b]

It is, of course, possible to combine several of these techniques in order to enhance the desired effects. One of the most effective hybrids is double-pane glass with an air layer between the glass sheets and a coating on surface of the inner pane. In Table 3-1 the performance of such a design is compared with that of standard plate glass under identical specified conditions. The double-pane design reflects nearly three times, transmits less than one-sixth, and conducts less than one-half as much incident energy as the single-pane design. Overall the relative interior heat gain due to energy passing through the single-pane window is about four times as great as that experienced with the double-pane coated window.

Such multiple-pane coated glass is obviously going to be more expensive than the single-pane variety. But the additional expense for the glass may be offset both by the possibility for using smaller building environmental systems which reduces first cost and by lowered fuel costs over the life of the building.

---

[b]This includes both the energy used directly in manufacture and the energy resources (e.g., petroleum) required as raw materials.

Table 3-1
Comparison of Relative Heat Gain through Single- and Double-Pane Windows

	Single, ¼ Inch Clear Plate Glass[a]	*Double-Pane Coated Surface of Inner Pane*[a]
Incident solar energy	230 Btu	230 Btu
Reflected	16	46
Reradiated and convected to exterior	28	140
Transmitted to interior	177	28
Reradiated and convected to interior	9	16
Heat conducted due to temperature differential	14	6
Relative interior heat gain	200	50
Percent of incident solar energy experienced as interior heat gain[b]	81%	19%

[a]An outdoor temperature of 89°F and an indoor temperature of 75°F are assumed.
[b]Does not include heat conducted due to the temperature differential.
Source: National Bureau of Standards, *Technical Options for Energy Conservation in Buildings* (Washington: U.S. Department of Commerce, July, 1973), pp. 80 and 85.

Consider the example of the Toledo Edison building, cited in a 1973 National Bureau of Standards study,

> The architects. . .selected a chromium coated dual wall insulating glass that made an added expense initially of $122,000 compared with the conventional ¼-inch plate glass. However, offsetting this initial expense was a savings of $123,000 in initial costs for the heating and cooling equipment, ductwork, and the like. . . . The payoff is the fact that the design chosen results in an energy consumption savings of 729.4 kilowatts per hour (translates to a savings in yearly operating costs of approximately $40,000).[5]

Storm Windows. Storm windows also reduce the solar transmission and the thermal conduction of windows and represent a high desirable energy conservation feature that can be retrofitted to most existing structures at a cost that will be paid back in fuel savings within a few years. Installation of storm windows reduces solar transmission by some 10–15 percent but may more than double the window area's resistance to cold weather conduction heat losses.[6] Actually the storm window concept itself may have more energy conserving potential than its current applications would indicate.

Since it is at times desirable to alter the solar gain characteristics of a window, it seems that a nultipane design which allowed for the movement (or perhaps removal) of one or more of the panes might give the desired flexibility. Individual panes might have different characteristics, such as differing thicknesses, coatings, or materials, thus allowing the alteration of the solar gain and/or conductive properties of the window unit as a whole by simply raising or lowering certain of these panes.

Window Frames. The framing materials and design are of considerable importance to the energy characteristics of the window as a whole. For example, aluminum window frames transmit nearly twice as high a percentage of the total heat transmitted through the entire window as do wooden frames. In summer (winter) aluminum frames are responsible for about 25 percent of the heat gain (loss) through the window. In addition, aluminum is typically manufactured by a highly energy intensive process and is accordingly even less desirable as a framing material. The combination of thermal performance and energy intensivity characteristics of aluminum make the popularity of aluminum framing for storm windows particularly disturbing, since it implies a significant offset to the energy conservation potential of such windows.

Window frames are thus of major significance to the loss/gain of heat. The choice of material, the area of the frame, and the tightness of its installation vis-à-vis the wall and the window panes are all important parameters.

Shading Devices. It is possible to reduce the intensity of solar heat (or light) gain through windows by using appropriate shading devices. Such devices may be placed either externally or internally and may be either fixed or movable. In general, external shading is significantly more effective. If properly designed, it can block between 20 percent and 30 percent more solar heat gain than high quality internal shading.

External shading can be accomplished by incorporating projections above (overhangs) and/or to the side (side fins) of windows into the building design. Such devices are fixed and must be placed with attention to the orientation of the window being shaded. Combinations of side fins and overhangs producing a kind of "eggcrate" pattern are sometimes required. Somewhat greater flexibility can be achieved by the use of movable awnings. However, awnings, whether movable or fixed, are far more sensitive to winds than solid overhangs or side fins. On the other hand, they are much easier to retrofit to existing buildings.

External shading may also be provided by other buildings or by trees, though the former is only a realistic design option when an entire complex of buildings is being constructed. Otherwise, the presence or absence of such effects must be taken as given. Trees, however, are generally more controllable. Unfortunately, they are typically usable only for shading up to a height of a few stories.

Even fixed external shading devices are capable of providing differential services at different times of the year. Whether in the northern or southern latitudes, the summer sun is high and the winter sun low. Overhangs thus tend to shade most effectively in the summertime and least effectively in the wintertime, which is exactly what is desired. Similarly, the use of deciduous trees provides for automatic adjustment of shading with the season. In the summertime when maximal shading is desirable, such trees are filled with leaves. As the weather turns cold, the trees lose their leaves and thus most of their shading ability, allowing interior solar gain to be enhanced just when it is needed.

Internal shading devices have the advantages of flexibility and essential immunity to the wear and tear of weather conditions. They are also relatively cheap and may be easily installed in existing as well as newly constructed buildings. In general, venetian blinds are somewhat less effective than roller shades in preventing solar gain. For example, using venetian blinds inside of a single pane plate glass window should reduce solar gain by between 36 percent and 45 percent, while roller shade can achieve between a 61 percent and 75 percent reduction (depending on whether they are translucent or opaque) under the same conditions.[7] However, venetian blinds have a very significant advantage as lighting control devices.

Again achievement of the optimum situation may require utilizing a combination of these devices, and again what is an optimal design for one location may not be optimal for another. However, it is clear that using shading devices of some sort will contribute to the reduction of negative energy flow in a great variety of cases.

Orientation and Extensiveness. During wintertime in the northern hemisphere, south-facing windows get direct sunlight through virtually all of the daylight hours. In the summertime, the height of the sun reduces this effect. In summer and winter, east-facing windows receive direct sun in the morning and west-facing windows in the afternoon, while windows oriented toward the north never experience direct sunlight. Thus the orientation of windows has important implications for both control of solar heat gain and lighting conditions.

Since the orientation of windows influences their utility as ventilating and aesthetic devices, it is not wise to offer generalized prescriptions for appropriate window orientation on the basis of lighting and heating considerations alone. Neither is it wise to ignore the fact that placing windows on a northern exposure mollifies the problem of summer solar heat gain, while placing them on a western exposure exacerbates it. Wind patterns, available view, internal orientation of building functions, sun exposure, and external temperature considerations all interact in determining the ultimate performance of windows, and they all must be taken carefully into account for the specific building site in order to optimize that performance.

It is, however, clear that the rather poor thermal performance of windows as compared to other building materials, should predispose us to avoid the use of vast expanses of glass. There is no reason why the aesthetic or lighting functions of windows should necessitate anything like the acres of glass that have become common in modern commercial buildings. The trend toward increasingly expansive use of glass in the skins of residential as well as commercial buildings should be reversed. It is a luxury which no longer makes economic, social, or ecological sense.

Doors

Like windows, doors represent thermal weak spots in the building's skin. However, unlike windows, the ability of doors to open and close is not merely desirable but absolutely crucial to the performance of their central design function. To be sure, we may fit certain types of doors with various devices in order to prevent their use at unauthorized times and/or by unauthorized persons. But the only rationale for having doors is to permit entry and egress. Protective and fire retardation functions would, in general, be more efficiently served by using solid walls. Of course, the entry and egress functions of doors could also be served by merely providing openings in the walls of appropriate size at appropriate locations. The use of doors rather than simple openings thus implies the desire for controlling the entry and egress.

When doors are opened, the potential for unimpeded, albeit temporary, contact between the external and internal environment occurs. Where the maintenance of a temperature (or humidity) differential between the two environments is desirable, such contact is to be avoided. Open doors allow heated indoor air to leak out and cold outdoor air to enter a building in winter and inversely in summer, thus requiring consumption of additional energy to restore the desired internal conditions. In the case of single family residential dwellings, the infrequency of openings and closings of exterior doors renders this effect relatively unimportant. However, in public, commercial, and industrial buildings a significant amount of energy can be lost in this way.

The simplest solution to this problem is the use of an airlock-type door. Probably the most common type of airlock door is the revolving door, which has the advantage of never permitting direct contact between the internal and external environments. Revolving doors can also be designed to collapse outward in such a manner as to allow more rapid exit under emergency conditions. In addition, standard doors may be provided for emergency use only. Alternatively, two standard doors separated by a small entry space can be used in place of a revolving door. The advantage of this type of airlock is that is facilitates the entry and exit of bulky objects, and accordingly may be the best configuration for a loading dock door. The disadvantage is that at times of relatively heavy traffic both double doors will be opened at once, thus destroying the airlock feature.

One or another configuration of the airlock door should be used for all doors, operated with reasonable frequency, which connect the building's interior with the exterior world. In the few cases where this may prove impractical (say, for example, in the case of a loading dock through which very large and bulky materials are transferred and in which space is at a premium), consideration should be given to the use of mechanized door closing devices, triggered either manually or automatically, which would facilitate the closing of the doors, giving some assurance that the doors would remain closed whenever feasible.

Because doors, virtually by definition, must be operable at the discretion of at least some (usually most) of the building's occupants, it is wise to provide design features that insure, to some extent, against the carelessness of those using the doors. Revolving doors are never open. Standard doors may be fitted with coils or may simply be tilted slightly so that gravity will close them. Ultimately, however, people can undo most of these design features so that there is real energy conservation value in making the closing of exterior doors standard operating procedure.

If doors are to be made of glass in order to function partially as windows, many of the energy considerations previously discussed for windows will apply to them as well. The orientation of doors relative to prevailing winds is especially important in cold climates whether or not the doors are made of glass.

Leaks

Although it is not a very elegant or sophisticated topic, no analysis of the prevention of energy loss could be complete without a discussion of the leakage problem. Heated or chilled air may leak through walls, windows, or doors. In fact, all buildings have a natural "breathing" rate, the rate at which air is exchanged between the external and internal environments without the requirement for conscious action by the building occupants.

Let me hasten to note that the exchange of air through the skin of a building as a part of its natural breathing is not inherently bad. If it were, the implication would be that all buildings should be as tightly sealed as possible, and we have argued very strongly in the preceding chapter that sealed buildings are often a prescription for energy disaster. But that is not to say that the breathing rate of a building should be determined by chance, or that is should be allowed to vary capriciously. On the contrary, the natural rate of infiltration and exfiltration of air is a design parameter important to the achivement of maximum net positive energy flow. As such, careful attention should be given to its determination, though with the understanding that the sensitivity of the breathing rate to construction practices prevents its precise determination by design alone. Maintenance procedures should correct alterations in this rate that occur due to building wear.

In other words, it is the variation in breathing rate from the intended level that should be a matter of concern. Whether due to building wear or accident or faulty construction, these excess passages of air exchange are the leaks that constitute an energy conservation problem.

Leakages wholly internal to the building can also produce increased energy consumption. For example, a leak in a steam pipe within an industrial facility can simultaneously add to the air conditioning load and require additional consumption of fuel in the generation of process steam. Hot water leaks can also dissipate inordinate amounts of energy. A leak of one drop per second will amount to a loss of approximately 700 gallons over the course of the year. If the water involved had been heated to 140°F from an initial temperature of 60°F, half a million BTUs would literally have gone down the drain from this one small leak.[c]

While it is difficult to quantify the leakage problem over the nation as a whole, it is clear that it represents a pure waste of energy—and one that is relatively easy to terminate.

The Control of Energy Exchange

There is no contradiction between the analysis and recommendations of this section and the earlier advocacy of design for interaction between building and environment. It is not the prevention of energy exchange that is advocated here, it is the *control* of that exchange. Insulation, multipane windows, shading devices, airlock doors, etc. are all control mechanisms that enhance the ability to manipulate energy flows—an ability that is critical to the minimization of building energy consumption. The use of some of or all these devices as tools for the maximization of net positive flow is thus completely compatible with, in fact is strongly indicated by, the energy conservative architectural philosophy and strategies discussed in the preceding chapter.

The Recovery and Recycling of Heat

Heat is generated as a by-product of the operation of all sorts of energy-using systems, biological, physical, and chemical. The heat is a by-product only in the sense that is does not directly contribute to the performance of the system's primary function. However it is hardly a by-product in the sense of being a small part of total product. On the contrary, it may be the largest part of the system's energy output. For example, heat accounts for 90 percent of the energy output of standard incandescent light bulbs yet contributes nothing to illumination.

A significant fraction of the energy generated intentionally as heat is also lost through dissipation after it has performed its primary function. Keeping to our objective of minimizing the need for newly injected energy implies that if we cannot avoid the generation of heat, we should recycle and reuse it as much as

---

[c]This does not include the energy required to purify and pump the water, only the energy used to heat it.

possible before it becomes so degrded that it is useless. In some cases, simple redistribution of the heat or control of where it is dissipated is sufficient to achieve very considerable energy savings, even when it cannot be reused.

Recovery of Heat from Light

All the sources commonly used for illumination also generate heat, whether fluorescent, incandescent, high pressure sodium, metal halide, or solar. However, this discussion will focus primarily on heat recovery from fluorescent lights for a number of reasons. First, they are the electric light sources most commonly used inside commercial, industrial, and other nonresidential buildings. Second, they are for the most part housed in groups in ceiling fixtures, a configuration that facilitates the direct recovery of heat. Third, the recovery and use of solar heat will be subsequently dealt with as a separate issue. Finally, casting the discussion in terms of fluorescent lights will allow specificity without much loss of generality—the systems described will work for any grouped fixture ceiling lighting system.

For any given level of illumination, the heat gain associated with fluorescent lighting is only slightly less than that associated with solar lighting.[8] Some 78 percent of the output energy of fluorescent bulbs is heat, and at that they are nearly twice as efficient as standard incandescent bulbs. In office buildings, heat generated by lights (primarily fluorescent) is responsible for 25 to 60 percent of the cooling load of air conditioning systems.[9] On the other hand, during periods when heating is necessary, efficient utilization of heat from light should reduce the load on heating systems significantly. In any case, it is clear that the proper management of heat-of-light is of considerable importance to the energy conservation problem.

Effective heat-of-light recovery systems do not require a particularly advanced or sophisticated technology. All that is essentially required is a system for the distribution and circulation of a heat absorbing fluid around or near the light source. The fluid absorbs heat from the lights and may then be used to transport the heat through insulated pipes or ducts to locations where heat is required. If an excess of heat is generated above needs, the fluid can be used to carry the excess to a point of efficient dissipation (e.g., a cooling tower), a point where it will not add to the load on other energy-using systems.

There are a number of such recovery systems currently available, some of which use air as the circulating fluid, while others use water. Typically, fluorescent light fixtures are hung from the true ceiling and set into a suspended ceiling, leaving a space called a *plenum* between the suspended ceiling and the floor above. Probably the simplest type of recovery system merely circulates air through this space, then draws the air away through a system of ducts. The net heat removal of such a system is about 20 to 30 percent.[10] A somewhat more sophisticated approach is to augment the effect of the air system by maintaining the pressure in the plenum cavity at a level slightly below that of the room itself. This pressure differential causes air to be drawn from the room, through the fixture, over the lamps, and into the plenum. Additional heat is thus picked up not only from the lamps but also from other heat sources in the room. Something between 35 and 45 percent of the light heat can be extracted by this type of design.[11] In water-cooled recovery systems, water passes through tubes which are part of the fixture itself. Typical entering water temperatures are in the range of 70 to 80°F. The net heat removal efficiency of a water-cooled heat-of-light recovery system can be as high as 60 to 75 percent, depending on entering water temperature and the rate of water flow. Supplementing the water effect by circulating air through the plenum can increase the heat recovery efficiency by as much as 10 percent.[12]

Water-cooled recovery systems are not only far more efficient than air systems at removing heat generated by the lights but also have several other advantages as well. Probably their greatest advantage lies in the fact that they require substantially less space than air systems. For example, a reduction of 37 feet was possible in the overall height of the 23-story Westinghouse Building at the Gateway Complex in Pittsburgh as a direct result of using water-cooled rather than air-cooled luminaires. In addition, 4,000 square feet of extra rentable floor space was freed up because of reductions in the size of duct shafts, and other related equipment.[13] Water systems also enhance the ability to store heat, are easily controlled, and may result in lower total operating costs.[14]

During summertime, it will be particularly valuable to have heat extraction capacity, as virtually all heat produced by lights will be considered a nuisance. However, even in summer there may be other functions within the building that can benefit from heat input. For example, hot water is required year round in bathrooms and in cafeteria kitchen facilities. Heat may also be required for various industrial or commercial purposes. Rudimentary heat exchanging devices can be used to transfer the heat from the water actually circulated in the heat-of-light system if, for one or another reason, that water cannot be used directly. Even when the heat so recovered is at too low a temperature to meet requirements, it can still be used to perform a preheating function. This will also reduce total energy consumption. Whatever heat cannot be reused in this way can be rejected from the building.

In winter, the heat generated by lights may be a valuable source of space heating at the point of its generation. Along with body heat and other by-product heat sources, a substantial contribution may be made to space heating, thus reducing the load on the heating system itself. During cold periods, the heat-of-light recovery system should only be used to remove heat from any areas in which by-product heat sources are generating more heat than is desirable and redistribute it to heat deficient areas. There may or may not be an energy advantage to be gained by redistributing the heat-of-light within heat deficient spaces, e.g., removing it from the top of the room and radiating it from baseboard units. But in no case should the system remove heat from areas that are already heat deficient in wintertime. This requires a system design that permits a redirection of water flow and a shut-off of parts of the system independent of other parts. Only such a system will have sufficient flexibility to minimize net energy consumption under the conditions of a seasonally changing climate.

It is important that lighting systems be designed to provide appropriate lighting conditions first. While credit should be taken for the heat *necessary* lights produce in the design of heating systems and heat-of-light recovery systems should be provided to remove and redistribute excessive light-generated heat, lighting systems should not be enlarged beyond optical requirements in order to double as heating systems. As will be subsequently discussed, electric space heating is inherently inefficient. And electric lights are much less efficient as heating devices than electric space heaters. Buildings that are designed to have all their heating needs met by lighting systems (e.g., the John Hancock Center and the Sears Tower, both in Chicago) must be designed with sufficient lights to provide heat for the coldest day, which will tend to be far in excess of optical needs. They also require that the lights be left on when it is not lighting but heating that is required. Such design is extremely energy inefficient.

Recovery of Heat from Other Sources

There are a number of by-product heat sources sommonly found inside buildings other than electric lights. For example, people generate a significant amount of heat as a by-product of their normal metabolic activity. A normal 154 pound male will produce 15 to 20 percent more heat energy than a 100 watt incandescent light bulb while at rest (seated) and nearly as much heat as four such bulbs when walking (at 3.5 miles per hour).[15] Machines, both of the office and industrial variety, also generate heat at rates that are subject to wide variation, depending upon their design and utilization. Industrial chemical processes may be either heat-generating (exothermic) or heat-absorbing (endothermic) and thus may play a role in the thermal situation.

The more concentrated and stationary a by-product heat source is, the easier it is to efficiently recover heat from it. Heat recovery systems for concentrated stationary heat sources may be designed along more or less the same lines as the heat-of-light recovery systems just discussed. The more mobile and dispersed the by-product heat source, the more we must rely on more generalized ventilation systems to pick up heat from these sources after they have dumped it into the air. Though this is somewhat less efficient, it is not at all difficult to accomplish.

Rotary heat exchanges (also known as heat wheels), heat pipes and "run-around" systems are all capable of recovering heat from exhaust air and transferring it to incoming air, thus preheating the intake air and reducing the energy required for space heating in cold weather.

Rotary heat exchangers consist of a wheel packed with material of high thermal conductivity placed so that half the wheel is in the exhaust air duct and half in the intake air duct. As the wheel rotates the portion that has been heated by the exhaust air stream comes into contact with and loses heat to the incoming air. The portion thus cooled rotates further and once again contacts the exhaust air stream. With an appropriate rotation speed this system may be quite effective. The wheel is often driven by a motor, because it is typically quite large. However, if it is possible to use a series of smaller wheels, it is conceivable to achieve rotation by virtue of the movement of the air.

A similar effect may be achieved by the use of banks of heat pipes. These pipes, made of a material such as copper, are filled with a refrigerant and sealed at both ends. Set permanently with one end in the exhaust duct and one end in the intake duct, the pipes transfer heat continuously between the two. Run-around systems consisting of two heat exchangers connected by a closed loop of liquid-filled pipe with one heat exchanger in each duct can also be used to recapture the heat in the exhaust air stream.[16]

Types of Heating Systems and their Energy Implications

Fossil Fuel versus Electric Resistance Heating System

Purely in terms of the useful energy output achieved per unit of energy consumed, fossil fuel heating systems, when properly maintained and adjusted, are far more efficient than electric resistance heaters. Even though the energy efficiency of electric heat *at the heater* is considerably greater than that of fossil fuel systems, the energy lost in the process of generating the electricity (and to a lesser extent in transmission and distribution) is so great as to reverse the ranking of electric heat efficiency when calculated from the point of initial energy consumption to the point of heat generation. This illustrates a broader and considerably more important engineering concept: *In general, the more transformations from one type of energy into another required by an energy-using system, the less energy efficient the overall system tends to be.* The reasons for this are rather straightforward. Since no energy conversion is ever 100 percent efficient, each time one form of energy is changed into another, some energy is lost. Thus when a series of such transformations take place, the overall efficiency is a multiplicative function of the individual efficiencies of each conversion. Because all of these multiplicative factors are less than 100 percent, their product is less than the least of them—perhaps considerably less. For example, the efficiency of an electric resistance space heater at the heater is about 95 percent. Being extremely generous and using an efficiency of 40 percent for the electric generating plant, the overall efficiency of electric heating would still be only 38 percent, far below the efficiencies of fossil fuel heaters which only involve one energy conversion.

Table 3-2 presents a comparison of rated and actually achieved efficiencies for fossil fuel and electric resistance heating systems. The differences between rated and actually achieved efficiency are partly due to the use of less than best available units and partly due to suboptimal (from an energy viewpoint) maintenance and operating procedures. In the case of electric resistance heating systems, the source of this deviation is almost exclusively in the electricity generation process rather than at the point of system operation and is apparently

Table 3-2
Comparison of Energy Efficiencies of Fossil Fuel and Electric Resistance Heating Systems

	Space Heating		
	Rated[a]	*Actual Commercial*[b]	*Actual Residential*[b]
Natural Gas	85%	77%	75%
Petroleum Products	80	76	63
Coal	70	70	55
Electric[c]	38	31	31
(at heater)	(95)	(95)	(95)
	Water Heating		
	Rated[a]	*Actual Commercial*[b]	*Actual Residential*[b]
Natural Gas	70%	64%	64%
Petroleum Products	55	50	50
Coal	70	70	15
Electric[c]	37	30	30
(at heater)	(92)	(92)	(92)

[a]Outer limit of efficiency of available units in substantial use. Electric generation efficiency is assumed to be 40 percent.

[b]Estimated average experience.

[c]These data do not include transmission and distribution losses.

Sources: Data on rated efficiencies of fossil fuel systems from Carol and John Steinhart, *Energy: Sources, Use, and Role in Human Affairs* (Belmont, Wadsworth Publishing Co., 1974) p. 54; estimates of actual efficiencies of fossil fuel systems and rated efficiency of electric system at heater from Stanford Research Institute, *Patterns of Energy Consumption in the United States*. (Washington: Office of Science and Technology, Executive Office of the President, January, 1972), pp. 153-154; figures for actual overall efficiency of electric resistance heaters calculated from data on electricity generating efficiency derived from Science Policy Research Division, Congressional Research Service, *Energy Facts* (Washington: Subcommittee on Energy of the Committee on Science and Astronautics, U.S. House of Representatives, November, 1973), pp. 33-34.

due mainly to the former effect. The former effect is also undoubtedly the major cause of whatever deviation there may be between rated and commercially achieved efficiencies for fossil fuel systems. However, it is clear that the differences between rated and actual efficiencies of residential fossil fuel heating systems are largely due to the operations and maintenance effect.

Commercial establishments nearly always use maintenance procedures far superior to those used in residences. This factor alone essentially explains the pattern of the differences between actual efficiencies given in Table 3-2. Since natural gas is easy to burn without sophisticated equipment and tends to produce little fouling of the system, very little maintenance is required. The tendency of petroleum products to clog nozzles and otherwise deteriorate the combustion efficiency of space heating systems is considerably greater than that for natural gas, requiring more maintenance. Coal is an even "dirtier" fuel, and thus maintenance of coal systems is more important still. Accordingly, the differences between residential and commercial efficiencies would be expected to be small for natural gas, larger for petroleum, and largest for coal heating systems.

The most striking and important conclusion to be drawn from Table 3-2 is that with the sole exception of coal-fueled water heating systems, fossil fuel heating systems are far more efficient than electric resistance systems, even where they are not exceedingly well maintained. If proper maintenance procedures are carried out, fossil fuel heating systems are at least 50 percent, and often more than 100 percent, more efficient than electric resistance heating systems without exception. These data have two major energy conservation implications: first, proper maintenance procedures are extremely important to the achievement of optimum energy efficiency in fossil fuel heating systems; and second, electric resistance heating systems are far too inefficient to be tolerated in widespread usage.

However, because of the differential availability of the various energy resources, one cannot merely choose a heating system on the basis of technical energy efficiency alone. For example, natural gas, the most efficient heating fuel available, is also the shortest in supply in the United States. On the other hand, coal, the least efficient heating system fossil fuel, is relatively abundant. The solution to this dilemma may lie in one of three directions: develop improved designs for systems that use the more abundant fuels; convert the more abundant fuels into a form in which they can be more efficiently used (e.g., coal gasification[d]); or consider entirely different heating systems.

There are two important potential advantages to using electricity rather than direct fossil fuels that should not be ignored. First, electricity may be generated by a considerable variety of techniques ranging from wind power and

---

[d]Great care must be taken to calculate the full energy implications of any such process to insure that the energy gains achievable by use of the higher quality fuel produced are not completely offset by the energy used in the conversion and transportation process.

hydropower through combustion of fossil fuels to nuclear fusion. Some energy sources (e.g., hydropower) cannot be efficiently tapped under existing and foreseeable technology without generating electricity as an intermediary energy source. Thus electricity generation allows for greater flexibility in fuel use. Second, electricity-using systems do not generate pollution at their point of use. Rather, the pollution with electricity occurs at its point of generation. If electricity is generated in large amounts at centralized locations, the source of pollution is more concentrated and thus pollution control measures will generally be more economical, more effective, and easier for the authorities to monitor.

On balance, the energy inefficiency of electric resistance heating is simply too large and important a consideration to be offset by the flexibility and potential pollution control advantages that are associated with electricity in most cases. But these advantages are sufficiently attractive to warrant close attention to the possibilities of using electricity for heating in some way other than in electric resistance heaters.

The Heat Pump

In the discussion of thermodynamics presented in Chapter 1, it was pointed out that heat tends naturally to flow from higher temperature sources to lower temperature sources and, further, that it will never spontaneously flow in the opposite direction (according to the second law of thermodynamics). However, just as water may be pumped up hill by performing work to overcome its natural tendency to flow in the other direction, so it is possible to pump heat "up hill" from a lower temperature source to a higher temperature source by performing mechanical work. A device that performs such a function is called a heat pump.

The heat pump takes advantage of two principles of physics: first, that a fluid absorbs heat in the process of changing from a liquid to a gaseous state (the heat of vaporization) and releases heat in the process of changing from a gaseous to a liquid state (the heat of condensation)[e]; and second, that it is possible to alter the temperature of a gas by changing its pressure and/or volume. The latter principle, the relationship between the pressure, volume, and temperature of a gas, is described by the so-called equation of state of an ideal gas, $PV = n\mathbf{R}T$. Here P, V, and T are the pressure (in atmospheres), the volume (in liters), and the temperature (in degrees Kelvin), respectively, while n is the mass of the gas (in moles), which is constant for any given gas, and $\mathbf{R}$ is the universal gas

---

[e]The amount of heat absorbed or released in any change that takes place at a constant pressure is called the *enthalpy change* of the system. Therefore, if a liquid evaporated at a constant pressure, the heat of vaporization would constitute a positive enthalpy change. Similarly, in changing from the gaseous to the liquid state under constant pressure, there is a negative enthalpy change for the fluid, measured by the heat of condensation.

constant (i.e., **R** is constant for all gases). Essentially then, for any given gas the equation of state means that the product of its pressure and its volume is equal to a constant times its temperature. All real gases basically behave according to this ideal gas law, except at low extremes of temperature and high extremes of pressure.

Depending on which direction the fluid in a heat pump is allowed to flow, heat may either be removed from a room and transferred to a hotter outside source (cooling cycle) or removed from a colder outside source and transferred into the room (heating cycle). During the cooling cycle, a cold liquid circulates through a coil inside the room. Since the liquid is colder than the room, heat naturally flows from the room to the coil thus cooling the room. As the liquid absorbs heat from the room it evaporates becoming a cool gas. The gas then passes through a compressor, which by performing mechanical work raises the pressure of the gas more than it reduces its volume.[f] Since the net effect of the compressor is to increase PV, it will cause the temperature of the gas to rise. The hot, high pressure gas now flows through an outdoor coil. Since the gas is hotter than the outside ambient temperature, heat naturally flows from the coil to the outside air. The fluid, still under high pressure, is cooled in this process but only to the ambient temperature. The warm, high pressure fluid is sprayed through a nozzle that causes its pressure to fall sharply, thus reducing its temperature considerably.[g] The cooled liquid-vapor passes through a condenser which reduces its volume while keeping its pressure essentially constant. Since V drops with P constant, T must be reduced. The fluid thus condenses into a liquid colder than the room temperature, which will accordingly draw heat out of the room as it passes through the indoor coil.

The heating cycle is essentially the opposite. A simple reversing valve directs the hot, high pressure gas from the compressor into the indoor coil. The coil, since it is hotter than the room, naturally loses heat to the room. The warm fluid is then sprayed through a nozzle, reducing its pressure and thus its temperature, and is further cooled by the condenser which reduces its volume at a constant temperature. The cold liquid is now colder than the outside air and therefore picks up heat from the outside as it flows through a coil. The heat it absorbs causes it to evaporate, and the resultant gas flows into the compressor. There its pressure is substantially increased, causing its temperature to rise. The now hot gas is ready to be passed once more through the indoor coil to heat the room.

The heat pump design and operation described is purposely highly generalized. Many of the specifics are omitted, and many refinements of design are pos-

[f] This may be achieved by allowing the gas to enter a cylinder and then driving a piston down in the cylinder thereby compressing the gas. The mechanical work done in moving the piston is thus translated into a temperature rise in the gas.

[g] The drop in temperature associated with a drop in pressure can be demonstrated rather simply. The air inside a tire that has been at rest for some time will be at the same temperature as the outside air. Yet if you let some air out of the tire, the air will feel cold. The air rushing out of the nozzle experiences a sharp pressure, and thus temperature, drop.

sible. But it does illustrate the general principle behind heat pump operation: change a fluid cold enough to draw heat from a cool source into a fluid hot enough to lose heat to a warm source and back again by changing its pressure and volume through the performance of mechanical work. It should also be clear from this discussion that the choice of working fluid is extremely important. A fluid must be chosen in accordance with the temperatures at which it undergoes phase changes and the heat absorbed or released during those changes.

The energy advantage of the heat pump as a heating system derives from the fact that the only energy it requires is that needed to operate its mechanical components (e.g., the compressor and pumps). It does not have to consume energy for the generation of the heat itself. Rather it uses the exterior world as its heat source in the heating cycle. Because it *concentrates existing heat* rather than generating new heat, it's contribution to the rate of entropy increase should be lower than that of systems that produce heat by chemical combustion or electric resistance. That must also be counted in its favor.

Heat pump efficiency is typically measured in terms of the coefficient of performance (COP), which is simply the ratio of heat energy provided to the energy consumed by the heat pump. Note that although this is an output/input energy ratio, it has a different character than the output/input ratio described in the preceding section as the efficiency of other heating systems. (cf. Table 3-2). The COP is the ratio of heat energy *transferred* by the heat pump to energy consumed, while the efficiency of a fossil fuel or electric resistance heater is the ratio of heat energy *generated* by the heater to the heater's energy input.[h] However, the COP is the efficiency measure most directly relevant to the evaluation of the heat pump as an alternative heating system.

By the nature of its operating principles, the COP of a heat pump will be a function of both the indoor and outdoor temperatures. Assuming a desired indoor temperature of 70°F, existing systems are capable of achieving a COP in the range of 2.5 to 3.5 when the outdoor temperature is as low as 45°F. As the temperature drops the COP drops. But even at an outdoor temperature of 10°F, the COP may still be between 1.5 and 2.0 Depending on the system design, the COP may not drop to 1.0 until the outside temperature falls as low as a few degrees above 0°F.[17] Thus, currently available systems are capable of supplying more than one unit of heat energy for every unit of energy they consume down to an outdoor temperature only slightly above 0°F.

This does not necessarily mean, however, that a heat pump is capable of supplying sufficient heat to maintain a 70°F internal temperature until the external temperature falls below 10°F. On the contrary, some supplementary heating may be required at outdoor temperatures as high as 40°F, though proper insulation and heat pump design can reduce this temperature to about 25°F.

---

[h]If we wished to measure the efficiency of a heat pump in a way directly comparable to the usual machine efficiency measures, we would have to take the ratio of the useful mechanical energy output of the pump to the pump's energy input.

The supplementary heat, which can technically be supplied by any alternative heating system, may not need to equal the heat energy supplied by the heat pump until the outdoor temperature falls 5 to 10°F below the temperature at which the supplementary system was first required.[18]

Because existing heat pump systems are typically electrically driven, the energy lost in the generation (and transmission and distribution) of the electricity must be considered in computing the overall energy efficiency of this system. Assuming an average 34 percent efficiency in the generation of electric power (and ignoring transmission and distribution losses[i]), the COP of a heat pump system must be nearly 1.5, for it to be as energy efficient as the least efficient fossil fuel space heating or water heating (except coal) system; to match the best average actual system performance, the COP would have to be about 2.2 (cf. Table 3-2). If the supplementary heating is provided by an electric resistance heating system (as is typically the practice), the COP would have to be that much higher over the range of temperatures at which supplementary heat is required in order to offset the inefficiency of that system. It is therefore questionable whether heating systems based on the heat pump will be as efficient as fossil fuel systems once the outdoor temperature drops much below 20°F. Certainly, they will not be as spectacularly efficient as the COP leads one to initially believe.

However, because they represent a relatively efficient way of using electricity as the direct energy source in a heating system, and because of the flexibility and pollution control advantages the use of electricity may afford, heat pumps do have real appeal. Certainly in climates not frequently subjected to temperatures below the mid-20s, heat pumps are a highly energy efficient means of providing necessary heat. Furthermore, the fact that the same unit can be used to cool as well as heat a building,[j] makes the economics of heat pumps more attractive. Still another advantage is the ability of a heat pump to use outdoor heat sources/sinks other than the air, e.g., the earth or groundwater, which may allow it to substitute renewable for nonrenewable energy resources and/or improve the system's energy efficiency.

Although the heat pump became commercially available during the late 1940s, maintenance problems plagued the earlier units and apparently created considerable consumer resistance. But vastly improved deisgns have made currently available units much more reliable and efficient. Their relative energy efficiency, particularly in milder climates, will make them an increasingly attractive alternative heating system.

---

[i]It should be noted that we have also ignored the energy in transporting fossil fuels in the calculation of the energy efficiency of those systems.

[j] The effect of falling outdoor temperatures on system efficiency during the heating cycle is logically symmetric to the effect of rising outdoor temperatures on the system's cooling efficiency.

Solar Heating

Without the use of exotic space-age photoelectric cells and the like, it is perfectly possible to use some of the enormous flood of solar energy that continuously pours over half the earth at a time, rain or shine, for direct space and/or water heating. Though solar heating systems may not be optimally efficient from a purely technical standpoint (as, for example, measured by the ratio of energy output to input energy), they do conserve nonrenewable energy resources and so are legitimately considered an energy conservation alternative. The added fact that they contribute almost no pollution makes them even more interesting.

Space Heating. Buildings may be designed to utilize incident solar energy for space heating by simply choosing proper construction materials, placing windows appropriately, etc., as has been previously discussed. But it is also possible to design specific systems for the heating of space by solar energy. Any such system must have three essential elements: a collector, a storage device, and a radiator. It may also have additional components, such as movable insulation, which increase its operating efficiency.

The specific design of a solar heating system depends strongly on such climatic and geographic factors as the diurnal temperature flux, the angle of incidence of sunlight, the number of days of cloud cover, etc. However, it is not difficult to outline a general design.

A solar collector, usually (but not necessarily) on the roof, absorbs incident solar energy which raises the temperature of a fluid in the collector. The heated fluid flows to an insulated storage device. Here either the fluid itself is stored or a heat exchanger is used to transfer the heat from the working fluid to a substance (solid or fluid) in the storage device. As heat is required inside the building an appropriate portion of either the collector fluid itself or the fluid from the storage device flows into the radiator system and loses heat to the building's interior. The intermediate thermal storage device is essential both to provide heating during nighttime and to allow the system to operate properly even when the cloud cover lowers the intensity of the solar radiation reaching the collector for more than brief periods. It cannot, of course, indefinitely compensate for such diminution in the solar flux.

Hay and Yellott describe a solar heating system built into a house in Phoenix, Arizona that is not only elegant for its simplicity but also can be used to cool the house during hot weather.[19] Ponds of water sealed in clear plastic lie on top of a layer of black plastic which lines a metal roof. A layer of insulation may be moved so as to either cover the ponds or expose them. And that is essentially the entire system!

In cold weather, the insulation is moved away during the daytime exposing the roof ponds to the sunlight. The sunlight heats the black plastic and the water, and a portion of this heat is conducted through the high thermal conductance metal roof to warm the interior. At night the insulation is moved over the roof to prevent radiation of heat to the night sky. The heat stored in the water radiates downward into the room. In hot weather, the insulation covers the roof during the day to prevent solar gain, and the roof water ponds draw heat out of the room cooling it. At night the insulation is moved away and the roof ponds radiate heat to the cool night sky. The water, thus cooled, can be used to air-condition the room when daylight returns.

This system was operated in Phoenix over a full year period during which ambient temperatures ranged from subfreezing to as high as 115°F. In the face of these outdoor temperatures, indoor temperatures were maintained between 68°F and 82°F without supplementary heating or cooling, except for a few days during December. During this abnormally cold and cloudy period, a simple 500 watt heater would have provided sufficient supplementary heat to maintain comfortable room temperatures.[20]

But solar heating systems are not only applicable to small houses in sunny places like Phoenix. A task force of building designers in New York City conceived of a hypothetical 43 story office building that would rely heavily on solar energy.[21] The building, which they called the Encon Building was to contain 1.1 million square feet of internal floor space. It was designed to be located on Third Avenue in Manhattan. The roof would slant toward the south at a 45° angle to enhance collection of solar radiation. Windows facing south and east would be shaded by projections that also serve as solar collectors. The design, as conceived, should utilize solar energy for 20 percent of the building's summer cooling and 30 percent of its winter heating requirements. It uses only presently available technology and is considered by its designers to be ready for construction.

Water Heating. Solar water heating systems may be designed along the same lines as solar space heaters. In such heaters, water is typically the working fluid circulated through the collector and stored in a well-insulated, appropriately sized tank. By virtue of thermosyphoning action, natural circulation of water (i.e., with little or no mechanical pumping assistance) between the solar collector and storage tank can be achieved, adding a small additional energy savings. However, this same thermosyphoning action will tend to cause the water to back-circulate in the evening and be cooled by the night sky. This effect will be essentially eliminated by setting the bottom of the storage tank at least 2 feet above top pipe of the solar collector and insulating the pipe carrying water from the top of the collector to the top of the storage tank.[22] If this is not practical in a particular installation, a pump may be necessary. There is, of course, no need for a radiator component in a solar water heating system.

For typical domestic hot water uses, a water temperature of 135°F is certainly sufficient. Global solar radiation patterns are such that over the most of the southern half of the United States a properly designed solar collector will only require an area of approximately 0.75 square feet per gallon to heat water to this temperature, provided that the rate of flow of water through the collector is set appropriately and the associated water tank is well-insulated. A family of four would need a tank of only 80 gallons and therefore a solar collector of about 60 square feet to provide hot water at 135°F in sufficient quantities for normal daily use.[23]

Thus in the southern half of the United States, solar water heaters could play a major role in reducing consumption of nonrenewable energy resources for water heating purposes. But what of the northern half? If the northern half of the contiguous United States is itself divided into two roughly equal parts, a northern and a southern section, then it turns out that the southern section of the northern United States experiences solar radiation only about 14 percent less (120 kcal/sq cm/ year), and the northernmost section experiences solar radiation only 29 percent less (100 kcal/sq cm/year) than the northern part of the south (140 kcal/sq cm/year).[24] These solar radiation figures are far too aggregated and far too generalized to be of use in a detailed evaluation of the feasibility of replacing conventional water heaters with solar devices. However, they do illustrate that the gross quantity of solar radiation received in the more northern latitudes of the United States is not so much lower than that received in the southern latitudes that the possibility of using solar water heaters there, at least as a supplement to conventional heaters, should be ruled out. It is an energy conservation measure that bears close scrutiny.

The Economics of Solar Heaters. The technology of solar heating systems for individual residences is so simple that it is not beyond the capability of an advanced do-it-yourself type to construct a reasonably efficient system—given attention to the available design literature or access to a set of basic blueprints. But this is not necessary.

There are a number of companies that sell solar water heating systems in the United States and elsewhere. There is, of course, considerable price variation, but solar collector units of up to 180 square feet were being sold in the United States in 1975 for prices between $170 and $225 per unit, depending on order quantity.[k] Entire systems for prices under a few hundred dollars are apparently available from foreign firms. Because of the simplicity of the technology involved, it is a virtual certainty that the production of such units is subject to very considerable economies of scale and will drop sharply when interest in the units becomes sufficient to sustain mass production.

---

[k]The higher price was for a single unit order.

Thermal Control in Design and Operation

Thermal Mass and Thermal Conductivity

Thermal mass is a term that describes a system's capacity to store heat, i.e., to serve as a heat reservoir. It is a property wholly separate from *thermal conductivity,* which refers to the ability of a material to transmit heat. Both of these properties are important energy parameters of building materials: the former is critical to the building's ability to smooth the external diurnal temperature fluctuations into more moderate internal fluctuations; the latter to the ability to maintain temperature differentials between the external and internal environments. Since the question of reducing thermal conductivity as a means to minimizing energy consumption in buildings has been dealt with in the previous discussion on insulation, we will focus primarily on the design implications of thermal mass here.

We have seen that differences in thermal mass are an important part of the primitive architectural response to climatic variations. But the thermal performance of primitive structures, though extremely impressive, does not always meet modern standards. Mechanical systems may be needed to supplement the natural thermal performance of the structure. Therefore it is necessary to consider the energy implications of thermal mass in buildings fitted with such systems.

High thermal mass structures are beneficial from an energy conservation standpoint during hot summer days. They reduce heat buildup in the building interior during the daylight hours by storing some of the incident heat in the building materials and reradiating it during the night. If the internal and solar heat gains are greater than the building's heat loss, high thermal mass may also be beneficial in the winter. This excess heat will be stored in the structure and can serve to maintain the internal temperature for a while after the daylight period ends and the various internal heat gains are diminished (because of people leaving the structure, machines being switched off, and so forth). Without high thermal mass serving as a heat reservoir, this excess heat would have to be rejected during the day and thus additional heat energy would have to be generated at night. However, high thermal mass is a disadvantage whenever mechanical heating or cooling systems are in operation.

Just as the sun's heat must first fill up the energy reservoir represented by a high thermal mass structure before it can heat the building's interior, so must the heat generated by a mechanical heating system. That is, the mechanical system may heat the air, but a portion of this heat will be absorbed into the walls and floors until they have been sufficiently heated to rise to the desired internal ambient temperature. Similarly, mechanical systems seeking to cool the interior must remove sufficient heat from the structural materials. Where mechanical heating or cooling systems are required, high thermal mass implies a greater heating or cooling load and thus greater energy consumption by those systems.

To a large extent, the thermal mass of a building is determined once it has been constructed, implying that the choice optimal thermal mass (from an energy conservation viewpoint) is made at the design stage. A whole series of local climatic, site, and other design factors determine the optimum thermal mass and must be jointly considered. But suppose the calculations indicate that one thermal mass is optimal part of the year and another thermal mass during another part of the year. What then?

It may be feasible to improve on the obvious possibility of choosing an intermediate thermal mass (which may never be optimal), if a design can be worked out to allow movement of a high thermal mass material inside or outside of the building shell as desired. In presenting this possibility, Stein offers the following example of how this has been accomplished: place one set of large water tanks inside the building and another set outside with connecting piping between; when high thermal mass is desired, pump the water into the inside tanks; when low thermal mass is preferable, pump the water from the inside to the outside tanks. By altering the position of the water storage, thermal mass is thus manipulated.[25]

Stein further points out that this system also offers the possibility of collecting heat on one side of the building skin and releasing it on the other. Thus daily cycling of some or all of the water can serve as a heat redistribution device. Of course, one must balance the energy savings attainable by such a system against the energy consumed by the pumping process, as well as the energy required to manufacture the components of this system. It is conceivable that there will be a greater net energy saving by moving the water only infrequently to alter the thermal mass than by cycling the water on a daily basis.

Internal Orientation of Functions

Placing functions that generate excessive heat as a by-product near the building perimeter facilitates dissipation of that heat, placing them in the central core of the interior facilitates reuse of that heat. To the extent that these uses involve heavy equipment or specialized facilities, they will not be readily moved and thus should be placed according to whether the heating or cooling season predominates. Again it is *net* energy savings we seek to maximize.

Under certain conditions it may be possible to save energy by grouping heavy energy users together and light energy users together. This will facilitate optimal sizing of building service systems as well as maximally efficient utilization of these systems, if the timing of the energy demands of heavy and light users is appropriate. For example, if the functions requiring more heating are grouped together and those requiring less heating are grouped together, then a high capacity heating system may be supplied for the former and a low capacity system for the latter. By splitting the heating system in this way, it is possible to shut off one part of the system completely when only one class of users requires heating and still achieve a high capacity utilization of the other part of the system without straining it. For this strategy to save energy, the timing of uses must be such that heavy users tend to operate together and light users tend to operate together, but both heavy and light users do not always operate simultaneously. Energy will be saved both because such a grouping facilitates shutdown of whole segments of the service system and because energy-using building service systems used near their designed capacity are almost always considerably more energy efficient than when there is a low capacity utilization.

A similar effect may be achievable, even where the timing of heavy and light energy uses is not so congenial, by simply designing the central part of the service system on a modular basis. For instance, several smaller boilers may be utilized in a heating system in place of one large boiler. This will increase energy efficiency only if the typical load is substantially less than total system capacity. As an example, the Stein New York City Public Schools study recommended a minimum of three boilers sized at 40, 40, and 20 percent of the total design on the basis of an Ohio State Engineering Experiment Station finding that the heating load was below 40 percent of design capacity 94 percent of the time in studies at one school.[26]

Energy may also be saved by grouping uses in accordance with their timing, even if the mix of heavy and light uses is not as previously discussed. If certain uses typically occur after hours, they should be grouped together so that services may be supplied only to a small part of the building during those periods. Of course, modular design of building systems, at least to the extent of being able to shut off services to unused areas, is necessary to the achievement of these savings.

Modular design of heating, cooling, and ventilating systems not only has an energy advantage in conjunction with properly grouped internal orientation of functions but also has the advantage of providing some back-up services in case of failures, without requiring installation of a completely redundant back-up system.

Thermostats

Thermostats are so important to the provision of appropriate internal temperatures with minimal energy consumption that there should be one thermostat for each separate functional space. This will avoid the necessity of supplying heating or cooling services at a specified level to an entire section of the building in order to achieve a desired condition in one part of that section. Plentiful thermostats will allow each area to contribute to the building's energy consumption only to the extent absolutely necessary to achieve the temperature level required within that area itself. In this way, no area need be excessively heated or cooled for the sake of another area's requirements.

Thermostats should be carefully placed within each area at a specific location that will permit them to get a true sample of the ambient temperature. This eliminates locations that may be exposed to direct sunlight or to the primary heating or cooling effects of any heat source or sink. At the same time, the thermostat must be readily accessible to room occupants authorized to operate it.

The temperature at which thermostats are set is of considerably importance to the energy requirements for temperature modification. For example, during the heating season in a typical U.S. climate, additional energy consumption will increase the heating cost by about 3 percent for every degree increase in the thermostat setting above 70°F.[27] The question of which thermostat settings are consonant with human thermal comfort is somewhat more complex than one might at first think and, for this reason, is dealt with in a subsequent separate section.

Aside from choosing a basic thermostat setting as low in winter and as high in summer as is consistent with thermal comfort, further energy savings may be achieved by providing for still lower settings in winter and higher in summer during periods of diminished activity. For example, setting the thermostat lower in the wintertime during the late night and early morning hours in residences or during nonworking hours in commercial and industrial facilities should reduce energy consumption. The degree to which energy is saved by such an operating procedure is determined partly by how long a period of "setback" there is and partly by the thermal mass of the structure.

The results of laboratory tests conducted by the National Bureau of Standards on a fully furnished, four-bedroom, 60,000 pound insulated wood-frame town house convey a feeling for the magnitude of savings achievable.[28] Setting back the thermostat from 75° to 65°F for an 8-hour night-time period produced a 9 percent heating energy reduction when the outside night-time low temperature fell to 3°F. Energy savings of 11.5 percent were achieved when the outdoor nighttime low and daytime high were 21°F and 65°F respectively. A computer study performed by Minneapolis-Honeywell in 1973 for 25 cities scattered around the U.S. predicted energy savings of between 7 and 12 percent for a 5°F nighttime setback and between 10 and 16 percent for a 10°F nighttime setback from a basic 75°F daytime setting.[29]

Water temperature settings are also typically excessive in the United States. Settings of 140 to 150°F and even higher are common in domestic hot water systems, whereas average hot water utilization temperatures are only in the range of 105 to 115°F. By way of illustration of the ubiquity of excessive hot water settings, the reader might consider how frequently he or she uses water at the maximum temperature at which it comes out of the hot water tap without first mixing it with cool water or allowing it to stand. The only residential use that could conceivably justify a water temperature of 140°F is dishwashing—and there is no reason why dish washing products could not be developed to work well with somewhat cooler (not necessarily cold) water. Extensive tests on the laundering of clothes performed by Consumers Union in 1974[30] indicated that the temperature of the water used made virtually no difference to the effectiveness of the laundering process.[l]

The National Bureau of Standards estimates that setting residential water thermostats to 120°F will save on the order of 12 to 15 percent of residential water heating energy by reducing the heat required to bring the water to its initial temperature and about 6 to 12 percent by reducing heat losses from storage tanks and piping.[31] The latter effect can also be accentuated by two design considerations: proper insulation of tanks and piping, and avoidance of unnecessarily long hot water piping runs. The length of hot water piping is particularly important because most residential uses of hot water are of relatively short duration, implying that a lot of hot water sits in the pipes and loses heat before it is ultimately used. This is the reason why it is necessary to let the water run out of the tap for a while before the water coming out of the tap becomes hot. Shortening hot water piping runs as much as possible, which involves attention to internal orientation as well as heating system design, will reduce the energy lost by dissipation of heat through the pipes.

Use of Underground Space

The high thermal mass of the ground is so efficient at smoothing diurnal and seasonal temperature fluctuations that the earth's temperature below the frost line varies very little for any given location.[m] For some purposes, this natural constancy of temperature can be a very desirable feature, contributing to our ability to maintain thermal control while minimizing energy consumption. It may therefore be beneficial to bury part of a building underground and assign the underground space to storage or computer facilities. Some precision manufacturing activities may also benefit from the constancy of temperature achievable with lowered energy consumption in underground facilities.

That underground storage is advantageous is certainly not a new idea (after all, wine cellars have been around for centuries), but it is another one of those old ideas that have been increasingly overlooked in the era of cheap energy and

[l] The formula of the detergent, however, did make a real difference. Interestingly enough, those detergents specially promoted for use in cold water did not perform as well in this study as general-use detergents containing sodium carbonate.

[m] For example, in the area of New Canaan, Connecticut the temperature of the ground below the frost line remains at about 52°F all year. See Carter B. Horsley, "Earth's Heat to Warm a Connecticut House." *New York Times,* January 12, 1975.

effective environmental modification machinery. Even where supplementary heating or cooling is necessary, underground storage is often energy conservative. In addition, this high thermal mass of the surrounding earth makes back-up cooling or heating systems largely unnecessary because internal temperatures will change only slowly even during extended periods of system shutdown. For example, in underground cold-storage facilities in Kansas City, the temperature rises only about 1°F per day during system shutdown as opposed to a rise of about 1°F per hour in similar above ground facilities.[32]

The constancy of underground temperature below the frost line makes the earth an excellent source or sink for a heat pump. Architect Landis Gores and heating consultant Paul Sturges designed a four-bedroom home in New Cannaan, Connecticut that utilizes the earth in this way.[33] More than half of the 3,300 square feet of usable floor space is below the earth's surface. A 2-foot air duct surrounds the house 8 feet under the ground and and is connected to a heat pump. Gores and Sturges estimate that the system will reduce consumption of fuel sufficiently to save about $1,000 per year in fuel costs (as against a $3,000 initial cost).

The need for constant mechanical ventilation must be balanced against the positive thermal properties of underground space. Underground space undoubtedly does have real energy conservation value for some specialized functions, but it is unlikely to be energy efficient (as well as psychologically acceptable) for many others.

Ventilation

Ventilation versus Circulation

Ventilation refers to the exchange of air between the inside and outside of a building. In contrast, *circulation* refers only to the movement of air and does not necessarily imply the exhausting of inside air and its replacement by outside air. It is important to differentiate between the two because they have very different energy implications. Circulation potentially requires energy consumption only to move the air; ventilation may require energy consumption to heat, cool, or otherwise condition the newly admitted air in addition to the energy required to move it. Unless the outside air naturally meets temperature and humidity requirements, ventilation adds to the load on the building's environmental service systems to a much greater degree than does circulation.

For some purposes, circulation is sufficient. For example, it it the movement of the air and not its freshness that aids the human body's ability to lose heat through perspiration. When circulation is sufficient to provide comfort, it is generally to be preferred to ventilation from an energy conservation standpoint.

Of course, a certain amount of ventilation is required for health and safety purposes. But typical standards for ventilation are far in excess of these basic requirements. For most sedentary uses, a ventilation rate of 5 cubic feet per minute (cfm) per occupant is satisfactory. Yet conventional practice calls for ventilation rates of 10 to 20 cfm per occupant.[34] Even a participant in fairly rigorous physical activity (e.g., in a gymnasium) does not require ventilation at a rate of 20 cfm for metabolic purposes—a rate of 15 cfm should be adequate.[35] A reduction of ventilation rates to the basic 5 cfm per occupant for most normal activities should result in a substantial energy savings. Since fans alone can account for 45 percent of total building electric power consumption, the approximately 30 percent reduction in fan load this would produce would result in a 10 to 15 percent electricity savings. In addition, the heat load could be reduced substantially by virtue of reduced introduction of outside air. For example, the Stein public school study estimated that a nearly 45 percent reduction in heat load was achievable by this closer alignment of ventilation rates with ventilation needs.[36]

In order to be able to replace some ventilation by recirculation of air, additional filtration of the air may be required. Since the degree of filtration required depends on the quantity of toxic and otherwise unpleasant substances and odors in the air, this tends to be more of a problem for industrial than for residential or commercial buildings. Filtration tends to add to energy consumption in at least two major ways: first, some energy is required to manufacture, transport, and maintain filters and filtering equipment; and second, the greater resistance to airflow due to filters installed in the air stream may require larger fans and ducts to cope with the greater load on equipment, especially as the filters begin to become saturated. These energy offsets must be balanced against the energy advantages derived from reduced intake of outside air.

Filtration and ventilation are substitutes chiefly in the areas of odor and toxicity control within the building's internal environment and not in the area of metabolic needs. Because sources of odor and toxicity are often (but certainly not always) localized, greater net energy savings may be achievable by supplying high ventilation rates at a few specific points rather than by elaborate filtering of large quantities of air. Ventilation rates in toilet areas of up to 50 cfm per fixture and use of exhaust hoods directly over cooking equipment and appropriate chemical laboratory and industrial processing equipment are examples. Metabolic needs can then be met by more generalized systems operating at ventilation rates governed primarily (if not solely) by those needs and supplemented by some degree of filtration.

Design for Natural Ventilation

Just as it is possible to conserve energy by management of solar gain, energy is also saved by replacing mechanical with natural ventilation. But proper provision of natural ventilation involves much more than simply providing operable windows. In order to effectively control ventilation with little or no mechanical assistance, one must understand the behavior of airflows over, around, and through buildings and their interior spaces. In the first place, it is not necessary, and sometimes not even advisable, to use operable windows as the means of admitting external or exhausting internal air. This may also be achieved by providing operable vents. In some cases, the separation of the lighting, view, and solar collection functions of windows from their ventilating function allows all of these functions to be performed more effectively. In other cases, safety considerations may militate against use of operable windows and so indicate the preferability of operable vents. In any case, whether windows or vents are used, some controllable openings are desirable wherever climatic conditions are reasonably changeable. In the following discussion, the term "window" is used as a general term for an opening through which air may be directed.

It is not only the size, shape, and position of windows that are important to effective natural ventilation but also their size, shape, and position relative to other windows and the size, shape, and relative position of barriers to the airflow (e.g., walls). Air flows from one place to another because of a pressure differential. It always naturally flows from high pressure to low pressure areas with its velocity and path determined by the pattern of pressure differentials and the position of obstructions.

When the wind blows against an abstruction like a building, an area of higher pressure is formed where the wind first encounters the obstacle. The air stream splits, flowing over and around the building, and the parts of the stream rejoin at some point *beyond* the far side of the building, but *not at* the far side wall. Thus a so-called "wind shadow," a low pressure area immediately behing the building, is created.[37]

Cross ventilation works to the extent that it provides a clear pathway for the air between the high pressure windward and the low pressure leeward areas. Therefore the openings must be oriented properly not only relative to each other but relative to the prevailing breeze. A cross ventilation path through a building set at right angles to the wind will have minimal effect. Furthermore the mere existence of a path for cross ventilation, even one that is properly oriented relative to the wind, does not guarantee that the airflow through the building will follow a pattern that will optimize natural ventilation.

The wind does not necessarily follow a direct, linear path from one window to another. For example, if windows were set at torso height on the windward wall and at the top of the leeward wall, one might expect the air flow to be largely directed upward along the ceiling, over the heads of the people in the room—and this would be likely to happen. However, adding a ceiling level outside overhang to the windward window, without changing the window positions, would result in an airflow directed *around* rather than over the people. Moving the outside overhang down to the top of the window would tend to reestablish the original up and over air flow, which would not be corrected by lowering the leeward window to the bottom of the far wall. In this latter case, the air would not flow directly from the torso height windward window down to the bottom leeward opening but rather up along the ceiling and down the leeward wall.[38] Thus it is of utmost importance to understand the dynamics of airflow in order to optimize natural ventilation. This is not to say a detailed knowledge of aerodynamic theory is required, but it is to say that a working knowledge of general aerodynamic principles is valuable. One cannot rely on simple-minded approaches or intuition.

Using natural ventilation for cooling in place of mechanized air conditioning devices is a highly effective way of conserving energy. Here the pattern and velocity of airflow are as important as its quantity, and these can be manipulated naturally by appropriate design. For example, the velocity of airflow may be increased (even above that of the initial wind) by reducing the size of windward and increasing the size of leeward openings.[n] Likewise, incorporating "pressure walls" (freestanding, air-directing partitions) into the interior design will allow a channeling of airflow around load-bearing or space-separating walls and into functional areas that might not otherwise receive natural ventilation.[39]

The cooling effects of natural ventilation may be enhanced by the use of supplemental mechanical devices. Although adding an exhaust fan here and there increases energy consumption over use of purely natural ventilation, it may allow the substitution of ventilation for air conditioning when natural ventilation is insufficient to achieve desired interior conditions. Thus providing supplemental mechanical ventilation may actually reduce the energy cost of achieving internal requirements. Similarly, simple cooling devices, like evaporative coolers,[o] can also be energy conservative. However, all such devices should be treated strictly as supplemental to natural ventilation.

Human Comfort

Heating, cooling, and ventilating systems are designed for the purpose of maintaining a comfortable internal building environment. When mechanical systems are employed for this purpose, energy is consumed—the greater the load on these systems, the greater their energy consumption. Therefore energy consumption may be minimized by designing on the edge of human comfort. For example, requiring heating systems to maintain the lowest comfortable internal temperature in winter will result in less energy consumption than requiring the tempera-

---

[n]This somewhat counter-intuitive statement results from the fact that using large windward and smaller leeward openings causes some of the incoming air to backup against interior walls, unable to rapidly exit, and thus produces a back pressure effect that reduces the airflow.

[o]An evaporative cooler consists of a container filled with absorbent material that is kept moist. As air passes through this container, possibly drawn by a fan, heat is transferred from the air to the water in the process of the water's evaporation.

ture to be held at the midrange of comfort. But what is the edge of human comfort?

Human thermal comfort depends on a number of factors: the amount and type of clothing, relative humidity, air velocity, degree of physical activity, mean radiant temperature, and dry bulb air temperature. The conditions under which individuals feel themselves to be thermally comfortable may also depend on age and sex, as well as on cultural and psychological factors. There is, of course, some degree of individual variation.

There have been a number of systematic inquiries into the question of human thermal comfort. One of the most complete was performed at Kansas State University.[40] Over a period of several years, 1,600 male and female college students were exposed, for a number of hourse each, to a uniformly heated or cooled environment with moderate air velocities and various relative humidities. Comfort was measured subjectively, with the subjects rating their impression of the environment on a seven point scale ranging from cold to comfortable to hot. All subjects were dressed in relatively light clothing whose insulating value had been predetermined. On the average, the subjects rated the environment "comfortable" over a range from about 76°F with 90 percent relative humidity to about 81°F with 15 percent relative humidity. However, the range of environmental conditions rated between "slightly cool" and "slightly warm" is much wider, going from roughly 68°F at 90 percent relative humidity to approximately 87°F at 15 percent relative humidity. In any case, it is clear that human comfort is consistent with a much wider range of humidity than temperature.

It is possible to get an idea of the influence of clothing on acceptable room temperature from calculations based on a theoretical comfort equation developed by P.O. Fanger.[41] The results, presented by the National Bureau of Standards,[42] indicate that at 50 percent relative humidity a sedentary adult would be comfortable at room temperatures ranging from 83°F (nude) to 74°F (business suit): at 100 percent relative humidity the range of acceptable temperatures would be 82°F (nude) to 70°F (business suit). If we focus only on the most relevant range of clothing–from shorts with an open neck shirt to a business suit–there is slightly more than a 4°F differential between the high and low acceptabe temperatures at 50 percent relative humidity (roughly 78°F and 74°F, respectively).

It should therefore be possible to achieve comfort at the range of temperatures classified as "slightly cool" to "slightly warm" in the Kansas State study by variation of clothing. If the building occupants are appropriately dressed, they may well be comfortable at an internal temperature as high as the mid-80s in summer (at say 20 to 30 percent relative humidity) and as low as the high 60s in winter (at about 60 to 70 percent relative humidity). Fortunately, it is not necessary to specify the outer bounds of this comfort range more precisely, since the internal building environmental systems are not to be preset and automatic. Rather, because building occupants will be able to alter the temperature and rates of ventilation within their individual space, they will do so in accordance with their own comfort needs. In order to avoid abuse of this control one need only set limits on the range of choice such that, say, most spaces where there is no vigorous physical activity (or other special considerations) cannot be cooled to below say 76°F in summer nor heated above, say, 72°F in winter (at appropriate humidities).

Considering the enormous energy-saving potential of setting temperatures higher in summer and lower in winter (several percent savings per degree), it is worth reevaluating our current patterns of dress. With the improvement of environmental modification systems, we have tended increasingly to dress according to our fancy rather than according to environmental conditions. For example, the midweight allweather business suit has more and more replaced the heavy winter-weight suit and light summer-weight suit combination. To some extent, the change to more uniform clothing weights has been forced upon us by the necessity of adapting our dress to the internal environmental conditions preset by automatic building systems over which building occupants have had little or no control. In many buildings wearing very heavy clothes in the winter and very light clothes in the summer is a guarantee of thermal discomfort for most people. It is not clear how much of this change in clothing patterns has been welcomed and how much has been forced upon the majority by the decisions of a few; but whatever the dominant cause has been, the effect has undisputably been a vast increase in the energy consumed by building systems.

By allowing building occupants to wear clothing more appropriate to climatic conditions, a considerable reduction in energy consumption is possible without sacrificing human comfort. But doing so implies the need for a cultural change. There is no technical reason, for example, why is should be taboo for business people to wear shorts or open neck shirts to work in the summer time. It will not addle their brains or make them lose their business skills. Nor will it necessarily lead to a reduction in dicipline or the integrity of the authority structure. If there must be differential clothing as a sign of differential status for cultural reasons, this can surely be achieved by differing styles and/or clothing accessories within the broad classes of light-weight and medium-weight clothing. All dress codes are essentially arbitrary, so there is no reason why we cannot settle on dress codes that make more energy sense than those that currently exist.[p]

Summary and Conclusions

A great deal of the energy currently used to heat, cool, and ventilate buildings can be saved without a significant sacrifice in human thermal or metabolic comfort. Liberal application of insulation to pipes and ducts carrying substances at temperatures different from desired ambient temperatures and to the walls, ceiling, and floors can reduce the load on heating and cooling systems consider-

[p]That is, if we must have dress codes at all.

ably at a fairly low cost. Insulation can be added to existing as well as new buildings. Excessive loss or gain of heat can also be prevented by the use of multiple-pane glass, with or without coatings, in windows shaded by external and/or internal devices and oriented appropriately relative to the sun. Some reduction in undesired heat loss/gain through fenestration in existing buildings can be achieved by use of storm windows and better internal shading devices. Airlock doors, oriented properly, and reasonable maintenance procedures involving prompt attention to leaks are also valuable in reducing the waste of energy.

Heat may be recovered and recycled from many sources that generate large aggregate amounts of heat as a by-product thus reducing the need for additional energy consumption to either generate heat to replace that which has been dissipated or to cool areas in which the heat from these sources has produced discomfort. Though heat is most efficiently recovered from fixed, concentrated sources like banks of electric lights, it is possible to recover some heat from much more mobile and dispersed sources, such as people. Essentially all heat recovery systems operate by circulating a heat absorbing fluid around the heat sources. In so doing, they consume some additional energy. However, on the whole such systems are likely to produce a considerable net energy savings.

Thermal mass, along with thermal conductivity, is one of the most important energy parameters of a building. The appropriate thermal mass for a building depends on climatic factors and thus cannot be universally predetermined. Thermal mass is generally set at the design stage, though it is possible to incorporate features into the building which will allow systematic alteration of its thermal mass during building occupancy. If a combination of high thermal mass and low thermal conductivity is required, this may be achieved, in some cases, with minimal use of materials (all of which require energy to manufacture and transport) by burying a part of the building.

Internal orientation of building functions has relevance to thermal control. Functions that generate excess heat should be located at the building perimeter if dissipation is desired and at the core if reuse is desired. Functions should also be arranged so as to avoid the necessity of piping heated or chilled substances (often water) over inordinately long distances. Thermostats should be provided for each functional space and should be subject to automatic setback (with manual override) during hours of low system load (e.g., nighttime, for residences).

Fossil fuel heating systems are considerably more energy efficient than electric resistance heaters and thus are to be preferred. However, electrically driven heat pumps may be even more energy efficient than fossil fuel heating systems and have the added advantage of doubling as cooling systems, thus increasing their economic attractiveness. With appropriate design, the energy efficiency of heat pumps should also be less sensitive to maintenance procedures than that of fossil fuel systems. The ecological implications of the large-scale replacement of fossil fuel systems by electric heat pumps are probably, on balance, favorable but require further serious study.

One heating system whose energy and ecological implications are clearly highly favorable is solar heating. The technology of solar heaters, contrary to popular belief, is extremely simple. Currently, systems are available at reasonably low cost and will become relatively less expensive as the prices of nonrenewable energy sources continue to rise and as the techniques of mass production are applied to their manufacture. Solar energy water and space heaters work well, and though the ratio of their output energy to their input energy is rather low for existing designs, the operating energy input is free, abundant, and nonpolluting. The use of solar heaters is not confined to intensely sunny areas. Most of the U.S. has sufficient incident solar radiation, rain or shine, to allow solar heaters to perform an important part, if not all, of the heating function.

Excessive ventilation wastes great amounts of energy, both because of fan requirements and because the exhausting of unnecessarily large quantities of heated or chilled air requires make-up heating or cooling energy expenditure. Current standards are excessive by a factor of two to four. Rather low rates of ventilation (on the order of 5 cubic feet per minute per occupant) are sufficient for metabolic purposes in most instances. Odor and toxicity control may be achievable at lower energy cost by a combination of high-spot ventilation rates and increased filtration of recirculated air.

Energy consumption by mechanical ventilating equipment can be drastically reduced by designing for natural ventilation. The use of operable windows of proper size and orientation and interior pressure walls can greatly enhance ventilation without large scale use of fans.

Human thermal comfort is consistent with a range of temperatures as wide as 15°F or more, if humidity, air velocity, and the type of clothing worn are appropriate. Since a reduction of several percent in the energy consumed by heating or cooling systems is achievable for every degree that thermostat settings are lowered in winter or raised in summer (within a reasonable range about the optimum), and since appropriate clothing can allow comfort to be achieved over a wider temperature range, it is clearly worthwhile to rethink our current patterns of dress. By rationalizing our dress codes with respect to climatic conditions, we can simultaneously increase comfort and energy savings.

Thus by a combination of increased insulation, improved window design, utilization of heat recovery systems, careful choice of thermal mass, thermal conductivity, and internal orientation of functions, plentiful use of thermostats, establishment of thermal and ventilation requirements at the edge of human comfort, use of underground space, and rationalization of patterns of dress, a very large net reduction in energy consumption is possible whatever heating system is used. Avoidance of electric resistance heating, much more extensive use of solar heating, and increased use of heat pumps where appropriate will also contribute greatly to the effort to conserve energy. Taken together these strategies will play a major role in achieving the estimated 50 percent reduction in building energy consumption previously discussed.

4 Lighting

Although lighting accounts for less than 2 percent of the total energy consumed in the United States,[1] it is responsible for nearly one-quarter of the energy consumed in the form of electricity.[2] In commerce, the fastest growing energy-use sector, lighting plays a particularly important role. Lighting systems in office buildings not only typically contribute at least 20 to 30 percent of the electricity demand directly but account for as much as 60 percent of the air conditioning load,[3] a load that is nearly always serviced by electrically driven machinery. On these grounds alone, lighting is an important point of focus for energy conservationists.

But the lighting situation is also important as a case study in the rampant expansion that has characterized the historical development of energy consumption in the United States. Furthermore, the lighting issue illustrates the extent to which we are still largely in the dark (no pun intended) on some fundamental questions of physiology and psychology relevant to the conservation of energy.

There are two basic physical units of particular importance to the discussion of lighting: the lumen and the foot candle. A *lumen* is a unit of the flow of light and is equal to the luminous flux radiating out from a uniform international standard point source of light, called the *candle,* through a unit solid angle (one steradian). A *foot candle* is a unit of illumination and is equal to the amount of direct light falling on a 1 square foot surface exactly 1 foot away from the international candle. Since 1 lumen flows through a spherical surface with an area of 1 square foot at a distance of 1 foot from an international candle, 1 footcandle is equivalent to 1 lumen per square foot. Thus the effective light output of a light source is measured in lumens, while the amount of illumination a surface receives from a light source is measured in footcandles (or lumens per square foot). For example, a standard 100 watt incandescent light bulb has an average light output over its life of 1,750 lumens; and accordingly, a 1 square foot area of surface 1 foot away from such a bulb would be illuminated at a level of 1,750 footcandles.

Clearly, the illumination that any surface receives depends not only on the strength of the light source but also on the distance between the source and the surface. However, the illumination does not diminish linearly with the distance but rather follows an inverse square law. That is, at 2 feet from the light source the illumination is one-quarter, not one-half, of what it was at 1 foot. Therefore, at 10 feet away from a standard 100 watt incandescent bulb a 1 square foot spherical surface receives only 17.5 (rather than 175) footcandles.

It should be borne in mind that although footcandles are defined in terms of spherical surfaces, most surfaces we need to illuminate (e.g., books, table tops, desks, balckboards, walls, floors, etc.) are typically flat. Thus 1 square foot of desk top 1 foot away from a 100 watt lightbulb will not receive a uniform illumination of 1,750 foot candles–rather only one point of that area will be illuminated at that level. The rest of the area will experience a lower level of illumination For small distances, flat surfaces of a given area are not as good an approximation to spherical surfaces as they are for large distances. But at large distances the illumination falling on a given surface area will be much lower than at small distances. Therefore there is a problem in providing uniform high levels of illumination.

Lighting Standards

The most frequently used standards of lighting are those promulgated by the Illuminating Engineering Society (IES), whose lighting handbook serves as the basic reference for lighting design. (The IES receives considerable input from the lighting and power industries.[4]) During the past four decades IES recommended levels for general interior illumination have increased more than 600 percent; since 1950 they have increased by about 250 percent.[5] This rise in lighting recommendations was, for example, reflected in the rising lighting standards for schools in the New York City public school system. Illumination levels recommended for classrooms were increased from 20 footcandles (fc) in 1952 to 60 fc in 1971; while 20 fc lighting levels were deemed sufficient for libraries in 1952, 70 fc were called for only 20 years later.[6]

Current lighting standards recommended by the IES specify minimum levels of illumination to be provided at every point on the work surface at any point in time. Because of the deterioration of the light output of all types of electric lights over time (due to dirt buildup as well as lamp depreciation), these standards actually require installation of sufficient lighting capacity to supply more than the specified footcandles initially–enough more so that at the point of maximum light deterioration prior to total failure the standards are still being met. Table 4–1 gives a few illustrative examples of present IES illumination standards.

The Steins have argued that since the IES handbook calls for the general lighting level to be designed for the most visually demanding task commonly found in an area and further specifies that the level of illumination in the workspace of the room be held to a reasonably close tolerance about the average footcandle level, these standards guarantee a uniformly high light level.[7] We have earlier discussed the consequences of this kind of worst case design for energy consumption with respect to ventilation and the load-bearing capabilities of structural members. The effect is no different here. Worst case design invariably leads to vastly increased energy consumption.

Table 4-1
Selected Minimum Illumination Levels Recommended by the Illuminating Engineering Society

Area or Activity	*Minimum Recommended Footcandles*
Residences	
Hallways, Conversational, and Recreational Areas	10
Reading and Study Areas	30–70
Kitchen and Work Shop Activities	50–70
Prolonged or Finely Detailed Sewing	100-200
Offices	
Interviewing, Washroom, and Inactive File Areas	30
Reading or Transcribing Handwriting in Ink or Pencil	70
Regular Office Work, Active Filing, Mail Sorting	100
Accounting, Bookkeeping, Business Machine Operation	150
Designing and Drafting	200

Source: Compiled from Illuminating Engineering Society, *IES Lighting Handbook* (New York: Illuminating Engineering Society, 1972).

The Blackwell Report

The very substantial rise in the IES recommended illumination levels over time has not been the result of purely arbitrary decisions. Rather, there has been a serious effort made to base these standards on scientific analyses of the relationship between illumination levels and the efficiency of performance of visual tasks. In 1959 the results of a body of research referred to as the Blackwell Report[a] were published.[8] The IES standards were subsequently increased substantially as a consequence of Blackwell's work, and his findings remain the scientific basis for the current high illumination recommendations.

Blackwell devised experiments to measure visual performance for both static and dynamic tasks. The static experiments involved the measurement of the subject's ability to perceive luminous discs projected on a uniformly illuminated screen. The contrast between the disc and the screen could be varied, but the position of the disc on the screen was always the same. The subject had only to indicate in which of four 2-second intervals the disc appeared. The dynamic experiments required the subject to indicate which of a series of discs mounted on the outer edge of a rotating wheel were lit. The subject could only see a part of the wheel at any point in time, so that each of the discs moved into and out of the visual field of the subject as the wheel turned.

The experimental data were used by Blackwell to develop a so-called *standard disc curve,* which, among other things, assumed an arbitrary disc size and a

[a]The research was supported by the Illuminating Engineering Research Institute, which is partly funded by the IES.

required 99 percent level of subject accuracy in correctly performing the experimental visual task. He then developed a device called a *Visual Task Evaluator* (VTE), which determines an equivalence between the contrast value of a real world task object (e.g., print on a page) and that of a luminous disc. The VTE-determined contrast value can then be located on the standard disc curve. Thus by combining the results of a VTE evaluation with the standard disc curve, the appropriate illumination level can be specified for any visual task.

Appel and MacKenzie have produced a brief but interesting critique of Blackwell's work.[9] They first point out that unlike the laboratory situation, the real-world contrast between an object and its background is independent of the level of illumination.[b] Secondly, they demonstrate that the arbitrary choice of 99 percent visual performance accuracy has very considerable influence on the resulting illumination recommendations. According to their calculations, requiring 99 percent accuracy results in recommended levels of illumination that are between 54 percent (for high-contrast tasks) and 233 percent (for low-contrast tasks) higher than those that would correspond to 95 percent accuracy and between 82 percent and 400 percent higher than those corresponding to 90 percent accuracy. Is a 4 percent increase in the accuracy of performance of artificially designed laboratory visual tasks worth a doubling or tripling of footcandle requirements?

Furthermore, Appel and MacKenzie make the point that many real-world visual tasks involve not merely vision but also touch, hearing, or smell. Probably one of the most common as well as most important examples of this is typing. An experienced typist does not require any significant level of illumination on the typewriter itself. Once his or her hands are on position on the keys few glances at the keyboard are required. Only the written material to be typed must be illuminated well, because it must be read. A high quality typist transcribing from a recorded tape can do so virtually in the dark. This is not to say that typists should be required to operate under such conditions, but it is to say that specifying lighting levels for tasks like typing on the assumption that they are purely visual is certain to lead to inordinately high illumination requirements.

Parameters Affecting Visual Task Performance Efficiency

The efficiency with which visual tasks are performed is not a simple function of the quantity of illumination. Rather, it depends on the interaction of a complex

---

[b]The contrast between an object and its background is given by: $C = (B_o - B_b)/B_b$, where B_o is the brightness of the object and B_b is the brightness of the background. But since the brightness of an object is given by $B = IR$, where I is the incident illumination and R the reflectance, $C = (R_o - R_b)/R_b$. Thus the contrast is not only independent of the level of illumination but is a constant for any given object-background combination.

of factors, some of which relate to the characteristics of the task object and some of which relate to the characteristics of the light provided. Of course, the visual capabilities of the subject are also quite important, but since this discussion is meant to be fairly general, normal human visual capabilities will be assumed. The effects of deviation from this norm will not be discussed. In addition, because it is the quantity of light that is most directly relevant to the issue of energy consumption, attention will be directed to the implications of variations in the other parameters for the level of illumination required.

Quantity of Light

It has been demonstrated that visual acuity increases as the level of illumination is raised over a very broad range under quite general conditions. There is evidence that the upper limit of this range may be beyond 1,000 fc in certain situations.[10] But whatever the precise value of the upper bound, it is well accepted that this increasing relationship is valid from very low to very high levels of illumination. The mere existence of a positive relationship, however, is insufficient for the proper evaluation of which light intensities are appropriate to the efficient performance of visual tasks. It is also necessary to know something about the sensitivity of visual acuity to variations in the quantity of illumination, as well as the significance of acuity gains to effective task performance.

Miles Tinker's analysis of the results of a quarter century of his own research and that of others into the lighting requirements for one of the most important visual tasks, reading, concludes:

> . . .visual acuity increases rapidly as light intensity is increased from a fraction of a fc to 5 fc and then gradually up to between 25–40 fc. As the illumination intensity is further increased up to 100 fc or even above 1,000 fc, the improvements in acuity are slight. Although the gains in acuity with intensities above 50 fc may have theoretical implications *they have no practical significance for reading* (emphasis added).[11]

Tinker further concludes:

> Gains in acuity, when illumination intensity is doubled, are very slight for levels above about 20fc. It is questionable whether the almost microscopic gains in visual acuity obtained under the relatively high fc of light justify their application to visual tasks such as reading. . . .[12]

It is also possible that excessively high levels of illumination will reduce the efficiency of performance of prolonged visual tasks, despite the slight gains in visual acuity they produce, by increasing eye fatigue. At any rate, it seems clear that illumination intensities in excess of 20 to 50 fc are simply not justified by any significant gains in the performance efficiency of reading or other similar visual tasks.

Degree of Patterning

Visual tasks differ in the extent to which the recognition of patterns in the task object is important to efficient performance. Reading a page of print is, for example, a highly patterned task, whereas reading the needle position on a finely calibrated gauge is not. The greater the degree of patterning in the task object the less it becomes necessary to see every detail of that object clearly and precisely. Rather, it is only necessary to see enough of the essential details to perceive the overall pattern.

This is the main reason why proofreading is such a difficult task to perform with very high accuracy. All of our training in reading is oriented toward improving our ability to recognize word patterns rapidly without studying every letter. Therefore, when we proofread we almost inevitably tend to overlook some minor typographical errors that do not severely disrupt the printed word patterns.

There are essential and nonessential details in every patterned task object. As long as we can see the essential details in the proper context, we can perceive the pattern. Alteration, absence, or inability to see the nonessential details does not significantly hinder task performance. By way of illustration, we typically do not find it difficult to read the account numbers printed on bank checks for machine reading, although there are many points of difference in appearance between those numbers and the numbers we normally see in print or handwriting. This is also why we have no particular difficulty in reading a wide variety of styles of lettering—they differ only in the nonessential details.[c]

One of the reasons why the Blackwell experiments are misleading is that they deal with completely unpatterned tasks, whereas the vast majority of real-world tasks involve some degree of patterning. They are therefore far more relevant to the setting of illumination levels for tasks like gauge reading and finely detailed industrial visual inspection than for reading, writing, etc.

In general, then, the higher the degree of patterning, the lower the level of illumination required to achieve high-efficiency visual task performance. Although the degree of patterning in the task object is typically an exogenous factor, under some circumstances one might conceivably be able to purposely increase the degree of patterning by an appropriate modification in the design of the task and thus permit the use of lower illumination intensities.[d]

---

[c]The dichotomy between essential and nonessential details is responsible for some of the difficulties experienced in learning to cope with different alphabets. We must learn which are the important and which are the unimportant details. For example, variations in the thickness of the stroke in a letter are important in Chinese but unimportant in English.

[d]Inspection for quality control is an area in which such manipulation of patterning is conceivable.

Size and Contrast

It is well understood that the ease of performing a visual task varies directly with both the size and contrast of the task object. It is easier to read large print than fine print, easier to read black print on a white page than on a dark brown page. Consequently, the illumination intensities required clearly vary inversely with object size and object-background contrast.

The tradeoff between size of the task object (or equivalently, its distance from the subject) and required lighting levels is assessed by the British Illuminating Engineering Society bulletin on school lighting as follows:

> . . .for the teacher to write on the chalkboard with letters 1½ inches high instead of 1 inch high would be as effective in approving visibility as would raising the level of illumination by 10 times. . . . For a child to move 4 feet closer to the chalkboard will have visual advantages. . . which can only be matched by raising the level of illumination by 30 times.[13]

This accords well with the experimental finding by Weston that a tenfold increase in illumination produced an improvement of 30 percent in the efficiency of performance of a simple visual task, whereas increasing the apparent size 10 times produced a 200 percent improvement in performance.[14]

Degree of Adaptation Allowed

The normal human eye is capable of adapting sufficiently well to establish visual comfort over a considerable range of light intensity. Pupil size is involuntarily reduced by contraction of the circular smooth muscle fibers of the iris in response to increased light or increased by contraction of the radial muscle fibers of the iris in response to diminished light. The visual sensitivity of the retina also adjusts to changes in the quantity of incident light. Adaptation to dim light may require up to 5 to 8 times as long as adaptation to bright light.[15]

In order to shed some light on the importance of adaptation to the efficiency of performance of visual tasks, Tinker performed an experiment in which the reading efficiency of subjects was evaluated for six levels of illumination ranging from 0.1 fc to 53.3 fc. When only 2 minutes of adaptation were permitted he found that light intensities below 10.3 fc significantly retarded the speed of reading, but with 15 minutes of adaptation the reading rate was not retarded until the level of illumination was reduced below 3.1 fc. In both cases, the speed of reading was the same for all the tested levels of illumination above and including the threshold values cited.[16]

In another interesting experiment with potentially far-reaching implications, Tinker allowed a group of subjects to adapt for 15 minutes to diffused light producing 8 fc of illumination. He then varied the light intensity and asked the subjects to choose which level of illumination they found preferable for reading.

Most chose 8 fc. He repeated the experiment allowing 15 minutes of adaptation to a level of 52 fc. Now the subjects most often chose 52 fc as their preferred level of illumination. From this Tinker concludes:

> Apparently, by picking an intensity and adapting the reader to it, one can obtain a preference for that intensity. If the investigator is interested in promoting use of lights of high intensity, the method of preferences will support it.[17]

If a visual task permits significant time for adaptation, there are at least some grounds for believing that it can be performed as efficiently at moderate levels of illumination as at high levels. It must be remembered that visual adaptation is required only when the intensity of illumination is altered. No matter how frequently or infrequently the visual task is performed, readaptation will not be required unless the person involved has experienced significantly different illumination levels in the interim.

In addition, to the extent that Tinker's results on the effect of adaptation on preferences are generalizable, they go a long way toward explaining why the higher levels of lighting recommended by the IES over time have produced so little consumer resistance and why so many of us now express preferences for lighting levels that would have seemed excessive only a few years ago.[e]

Quality of Light

The same quantity of light can have very different implications for the efficiency of performance of any given visual task depending upon characteristics of that light. The term *quality of light* refers to the mix of these characteristics taken as a group. Among the factors considered are harshness or glare and the spectral properties of the light.

Harshness describes the degree of concentration of the light source. In general a well-diffused (and therefore soft) light produces a more pleasant and less fatiguing visual environment than a concentrated light. This is, for example, why so-called soft white frosted incandescent light bulbs produce a more comfortable light level than perfectly clear incandescent bulbs of the same lumen output. This is also why many light fixtures have a diffusing globe or panel designed into them.

Highly reflective surfaces tend to reflect incident light without diffusing it and therefore create harsh points of light as compared to less reflective surround-

[e] I believe that the importance of adaptation to establishment of preferences applies far more generally than even Tinker implies. It seems to be operative with respect to virtually all the sensory stimuli that impinge on us. The difficulty that persons accustomed to quiet areas experience when first trying to sleep in the "big city" and the parallel difficulties that manv city dwellers experience before adapting to the deafening silence of less populated areas are nonvisual examples. Yet another is the tendency of people to acquire a taste for foods that they do not initially enjoy, upon repeated exposure.

ing surfaces. They may even concentrate light, augmenting this effect. Reflected glare is not only visually unpleasant and fatigue inducing but may readily obscure the task object by, in effect, overwhelming the eye and thus reducing the apparent contrast between that object and its background. Of course, for any given reflectance of the task object and its background, there is a direct, positive relationship between the amount of reflected glare and the harshness of the primary light source.

The spectral properties of light are important to both color perception and visual sensitivity. Visible light is a very small part of the electromagnetic spectrum, consisting of energy associated with wavelengths ranging from about 380–760 nanometers (i.e., billionths of a meter). Within that range, differences in wavelength are perceived as differences in color. In order of increasing wavelength, the visible spectrum consists of violet, indigo, blue, green, yellow, orange, and red. Other colors are the result of the interpretation of combinations of different wavelengths. Perfectly white light consists of all visible wavelengths in equal quantities.

The color attributed to any given object is partly a property of the object and partly a property of the light that illuminates it. All objects that are not in themselves light sources are perceived by virtue of the light they reflect. Since most objects do not reflect all wavelengths equally, they appear to have the color of the wavelengths they do reflect. For example, an object whose surface reflects light primarily ranging between 600–620 nanometers in wavelength will appear orange. But clearly in order for that object to reflect light in that range of wavelengths and thus appear orange, the light that illuminates it must contain those wavelengths. If it does not, the object will not appear orange but rather will appear to be the color corresponding to whatever other wavelengths present in the incident light it is capable of reflecting.[f] If it can only reflect light of 600–620 nanometers wavelength and the incident light is completely lacking in those wavelengths, it will appear perfectly black no matter how intense the illumination.

A light is said to be of higher quality, in terms of its color-rendering ability, the greater the extent to which objects illuminated by it appear to be their true color. Because color is a function of the illuminating light as well as the object's own reflective properties, true color must accordingly be defined with respect to some specified type of illumination. Typically, we define *true color* as the color that the objects appears to be in daylight, a light that contains all the visible wavelengths in approximately equal strength. Thus an artificial light source that does not radiate a full continuous spectrum will not render all colors truly.

A glowing tungsten filament produces light containing a continuous spectrum, whereas a mercury arc has a line spectrum. This is why objects look essen-

[f] Since the object must be less efficient at reflecting those other wavelengths (or else it would not appear orange when full spectrum light is shone upon it), the object will also not appear to be as bright.

tially normal color when illuminated by standard incandescent bulbs but look strange under a mercury vapor street lamp. For similar reasons, fluorescent lamps do not render color as well as incandescent lamps.

The spectral qualities of light are also important to general visual perception, because the human eye is not equally sensitive to all wavelengths. On the basis of considerable empirical observation, a *spectral luminous efficiency* curve has been established for the average eye, relating the relative perceived luminance of equal amounts of energy to the wavelengths associated with that energy. The curve, which has the appearance of a normal probability distribution, peaks at 555 nanometers wavelength (yellow-green) for the light-adapted eye.[18] From this curve one may readily determine the relative amounts of energy at different wavelengths required to yield the same perceived luminance. For example, approximately nine units of energy at 650 nanometers (red) are required to produce the same visual brightness response as one and two-thirds units of energy at 600 nanometers (orange-yellow) or one unit of energy at the peak efficiency 555 nanometers wavelength.

It might at first seem that considerable energy could be saved by choosing light sources that concentrate their energy in the central portion of the spectral luminous efficiency curve. In that way, equivalent levels of perceived illumination could be achieved with greatly reduced energy. However, there are at least two reasons why this approach may not be optimal. First, unless the objects involved in the visual task are relatively efficient reflectors of wavelengths in the modal range of visual sensitivity they will not appear to be illuminated by that light. Thus only objects whose true color is either white or one of the wavelengths in that range will be seen well. Second, it is conceivable that prolonged exposure to light of highly circumscribed wavelengths could produce deleterious physiological or psychological effects (including psychosomatic ailments).

Nonvisual Correlates

It was earlier pointed out that many tasks that we tend to think of as being visual are actually multisensory and, in fact, may often depend primarily on senses other than sight for the experienced performer. Typing, knitting, playing the piano, and some types of sewing are all examples. There are, of course, some classes of tasks that are purely or highly visual, even for the most expert individuals—probably the best example of such a task is reading.[g]

Whenever a task is multisensory, the environmental requirements relating to all the engaged senses should be considered jointly, and a comfortable perfor-

[g]Note that even purely visual tasks may sometimes be altered so as to eliminate the visual component—for example, reading by the braille system.

mance environment established. To the extent that it is possible to trade off other senses against sight in task performance, it may be possible to reduce lighting requirements in some circumstances. For example, if a task requiring the performer to precisely position one object relative to another is designed so that there will be an audible click when the exact position desired is achieved, the lighting levels will not have to be as intense as if the entire positioning must be done visually. We may also seek to redesign visual tasks to reduce the visual acuity and thus the level of illumination required for efficient performance, e.g., by replacing a gauge designed with a thin needle moving against a semicircular background of finely calibrated thin lines with a gauge consisting of a linear color bar moving horizontally against a background of black vertical calibrations.[h] At any rate, if the total sensory character of the task is taken into account, the result will tend to be less extreme lighting requirements than if we persist in assuming the task is purely visual. Again, of course, any redesign of the task must be carefully evaluated in terms of its implications for safety, efficiency, and net energy requirements from all sources and not merely the effect of the redesign on lighting energy needs.

Time Factors

There are two ways in which time enters into the performance of visual tasks: first, there may be a specific maximum amount of time allowed for any single performance of the unit task; and second, it may be necessary to repeatedly perform the task over a shorter or longer time period. If the unit task must be performed more quickly, the task object must be perceived more precisely with less adaptation. This would seem to imply higher required levels of illumination, and this is often the case. It is also possible that task redesign will yield greater benefits than a simplistic massive footcandle increase. However, the illumination implications of extending the work period over which the task is repeatedly performed are not as clear.

For years, the lighting industry has been conveying the message that increased illumination reduces eye fatigue, the logic essentially being that if more light is available the eye has to strain less to see something and thus will become less tired. But that is not so obviously sensible as it sounds. When the light intensity increases, the contraction of the circular smooth muscle fibers of the iris reduces the size of the pupillary opening and sensitivity reducing photochemical changes take place in the retina, as previously mentioned.[i] It is not at all clear

[h] Automobile speedometers were redesigned essentially in this way.

[i] These changes imply that the increase in effective light striking the retina are not as great as the increase in the lighting intensity.

that these changes, along with the higher level of excitation of the fibers of the optic nerve and the corresponding increased reception of impulses by the brain implied by a higher level of incoming visual sensations, necessarily reduce the fatigue associated with vision over as wide a range as seems to be suggested by the level of the current IES illumination standards. It may well be that prolonged exposure to illumination levels of 100 fc, 150 fc, or even 200 fc may increase rather than reduce eye fatigue. At any rate, what is obviously required is further research into the question of eye fatigue and not a blind acceptance of this lighting dogma. It is a vital question because the measure of the efficiency of performance of visual tasks relevant to the real world is not the accuracy of performance of highly simplified unit tasks over a short period but rather the accuracy of performance of real-world visual tasks over the total periods of time during which they are typically performed. Even 99 percent accuracy in performance of tasks for 30 minutes that are normally performed over 4-hour stretches means little, unless we can be sure that eye fatigue does not cause efficiency to drop off sharply after that initial fresh period.

Toward More Realistic Lighting Standards

It should be clear that the current lighting standards promulgated by the Illuminating Engineering Society are not as firmly based on scientific analyses accepted as relevant to the performance of actual visual tasks under real-world conditions as one might initially believe. Rather, there are very considerable grounds for believing that these recommendations are, in fact, excessive for the vast majority of cases. The energy consumption implications of this situation are quite disturbing, because the IES standards are widely adhered to as operational *minima* for the design of lighting systems throughout the United States.

But, if the current IES standards are excessive, just how excessive are they? Though a detailed answer to this complex question requires considerable scientific analysis, the data presented in Table 4-2 are intended to convey some feeling for the gross differences between the current IES standards in the United States and alternative feasible standards. The data are admittedly quite limited, dealing only with general levels of interior illumination recommended for selected areas in school buildings. Nevertheless, it is interesting to note the relatively close correspondence between the illumination recommendations of the British Illuminating Engineering Society, the New York City Health Code, and the standards that Tinker concluded were sufficient after a detailed analysis of lighting research, as contrasted with those of the United States IES. In virtually every case, the United States standards are 2 to 5 times as high as any of the others. It would seem likely, then, that a 50 to 60 percent cut in present United States illumination recommendations would produce lighting levels far more closely attuned to actual illumination needs than those currently provided.

Table 4-2
Comparison of Illumination Recommendations for Schools

Areas	*IES (USA)*[a]	*IES (Britain)*[b]	*NYC Health Code*[c]	*Tinker*[d]
Classroom	70–150 fc	20–30 fc	30 fc	20–30 fc
Library Reading	30–70	30	30	15–35
Office	70–150	30	–	15–25
Drafting/Sewing	100–150	70	50	40
Washroom/Locker	20–30	7–10	10	–
Laboratory	100	30	–	–

[a]Derived from recommendations in the 1972 *Lighting Handbook* of the Illuminating Engineering Society of the United States.
[b]Derived from the 1961 lighting code of the British Illuminating Engineering Society.
[c]Derived from the 1959/1964 New York City Health Code.
[d]Derived from the recommendations of Miles A. Tinker.
Sources: Adapted from Richard G. Stein and Carl Stein, *Research, Design, Construction, and Evaluation of a Low Energy Utilization School* (Washington: National Science Foundation, August, 1974), pp. h-2/1 and h-2/2; Miles A. Tinker, *Bases for Effective Reading* (Minneapolis: University of Minnesota, 1965), pp. 233–234.

In addition to drastically reducing the required levels of overall illumination, considerable additional energy savings should be possible by loosening the present uniformity prescriptions that lead to worst case design. Stein points out that there has been neither physiological nor psychological verification of the IES assumption that lighting levels must be kept within 25 percent of the required intensity and that, in fact, human eyes are normally able to adapt well to illumination intensities as sharply contrasting as sun on the snow or the shadows of a tree's branches (1,500 fc versus 20 fc).[19] It is also possible that excessive uniformity of light levels may actually be deleterious, producing a visually monotonous and unstimulating environment. By abandoning the excessive uniformity that leads to worst case design, lighting systems may be produced that provide sufficient specific lighting on the task object and in its immediate vicinity whenever the task is actually being performed, along with a considerably lower level of general area illumination. This will allow very considerable energy savings over present practice and will simultaneously correspond much more closely to actual visual requirements.

Use of higher reflectance ceilings and lighter colors for walls, floors, and furnishings in locations where visual tasks are important can also be energy conservative, directing light more completely to useful areas. Realistic lighting standards would specify such reflectance requirements as well as taking into account all the other parameters relevant to visual efficiency discussed in the previous sections. In this way, lighting levels could be tailored to specific needs, allowing a considerable reduction in wastage of lighting energy.

Artificial Light Sources

Although it is well-known that light can be produced by both combustive and noncombustive chemical reactions, virtually all artificial light sources currently in use depend instead on the light generating effects of electron flows. At the present state of technological development these electrically energized light sources have considerable advantages in terms of a combination of economic, safety, and performance efficiency factors. Yet there is a very considerable variation in these factors among the many existing alternative designs of electric light sources themselves.

There are three main categories of presently available designs: incandescent, fluorescent, and high-intensity discharge (HID).[20] Incandescent lamps produce light by virtue of a filament heated to incandescence by the flow of an electric current. The filament, typically made of tungsten, is encased in a sealed glass bulb that is either evacuated or filled with an inert gas, in order to prevent the rapid disintegration it would experience if exposed to oxygen. The use of inert gas tends to reduce the rate of evaporation of the filament and thus prolong lamp life. The inside of the glass bulb may be frosted by a light acid etching in order to diffuse the light. Glare may be further reduced by coating the inside surface of the bulb with white silica, but this introduces about a 2 percent penalty through the absorption of light by the coating—a penalty not measurably incurred by the frosting procedure.

Light is produced in fluorescent lamps by a somewhat more sophisticated process. The lamp consists of a glass tube filled with mercury vapor at low pressure and one or more inert gases (e.g., argon). An electrode is sealed in each end of the tube, and the inner walls of the glass are coated with fluorescent powders (e.g., zinc silicate, calcium halo phosphate, magnesium tungstate). Applying an electrical potential (i.e., voltage) to the electrodes readily ionizes the inert gas causing a current to flow between the electrodes. The electrons collide with the mercury atoms temporarily altering their atomic structure. As the disturbed atoms return to normal, they emit energy in the form of electromagnetic radiation of primarily 253 nanometers wavelength (ultraviolet). This radiation strikes the fluorescent powders, which absorb it, then reradiate the energy at wavelengths increased enough to cause them to fall within the range of the visible spectrum. Thus a fluorescent lamp transforms electrical energy into untraviolet radiation and then ultimately into visible light.

High-intensity discharge lamps operate in a manner very similar to that of fluorescent lamps. The chemicals enclosed in the tube may be mercury and argon (mercury vapor lamp); mercury, xenon, and sodium (high-pressure sodium lamp); or mercury, argon, and iodine compounds of metals like indium, scandium, sodium, thallium, etc. (metal halide lamps). However, in order to highlight the differences between HID and fluorescent lamps, we will focus on the most similar, the mercury vapor lamp.

In mercury vapor lamps the bulb containing the electrodes, the argon, and the mercury is made of quartz, not glass. The application of a voltage to the electrodes ionizes the argon easily thus initiating electrical arc. The heat of the arc vaporizes the mercury (which is visible as droplets of liquid mercury in an unlighted bulb) and generates a mercury arc that produces a spectrum of radiation with strong lines in the ultraviolet and visible regions. An outer bulb, made of glass and filled with inert gas, shields the quartz tube containing the arc from temperature changes and cuts off virtually all of the ultraviolet radiation, allowing only the visible wavelengths through.[j] The inner surface of the outer bulb may be coated with a phosphor that absorbs the ultraviolet radiation and emits visible light, improving the efficiency of the lamp (the mercury-fluorescent lamp). The mercury HID lamp operates at a much higher vapor pressure than does the fluorescent lamp and most of the visible light it generates, even in the mercury fluorescent lamp, is produced at the mercury arc. Whereas over 90 percent of the light emitted by a fluorescent light is the result of fluorescing of the phosphors, at least 90 percent of the light produced by the mercury-fluorescent is arc emitted.

Comparative Energy Efficacies

The efficacy of an artificial light source is measured by the illumination it produces relative to the energy it consumes. The usual measure employed is lumens per watt. There are two considerations that must be borne in mind when evaluating alternative light sources in terms of their efficacy: first, the wattage input does not truly measure the energy consumed because there are large energy losses in the generation, transmission, and distribution of electricity; and second, a great deal of the energy output from an artificial light source is in the form of heat not light, and this heat may either serve to reduce heating loads (a plus) or to add to cooling loads (a minus). The first consideration implies a multiplier effect common to all savings of electricity—a saving of one unit of electrical energy is approximately equivalent to a saving of three units of energy. However, it is not directly relevant to the comparison of artificial light sources so long as all sources considered are electrically energized. The second consideration is important to the total design and coordination of building energy systems, but it must be uniformly looked upon as a disadvantage in the design of artificial light sources. We may be able to lessen the disadvantage implied by the generation of this heat by reusing it, but since we know that electric radiant heating is inherently energy inefficient, it would be far better to develop artificial lights

[j]This latter function is extremely important because excessive ultraviolet radiation is very dangerous to the eyes and to the skin. Mercury vapor lamps whose outer bulbs are damaged should be immediately replaced.

that have an increased proportion of their energy output in the form of light rather than heat.

The efficacies of the three categories of electric lamps that have been discussed are compared in Table 4–3 by reference to selected specific lamps within each category. It is clear that incandescent lamps (whose energy output is approximately 90 percent heat and 10 percent light) are by far the least energy efficient, with efficacies ranging from 11 to 22 lumens per watt. Fluorescent lamps are considerably more efficient, with efficacies ranging from 50 to 73 lumens per watt, while high-intensity discharge lamps are generally still more efficient, outputting from 46 to 100 lumens per watt. Yet even the most efficient HID lamp in Table 4–3 does not approach the theoretical maxima shown.

The comparative energy inefficiency of incandescent lamps is strikingly illustrated by the fact that a 400 watt metal halide HID lamp generates 36 percent more total lumens with 60 percent less energy than does a 1,000 watt general service incandescent lamp. Similarly, two 24-inch fluorescent lamps produce about the same order of total illumination as a 100 watt general service incandescent lamp but do so at nearly three times the efficacy.

Table 4–3
Energy Efficacies of Selected Artificial Light Sources and Theoretical Maxima

Source	*Approximate Lumens per Watt*
Incandescent	
40-watt general service	11.0
60-watt general service	14.3
100-watt general service	17.4
1,000-watt general service	22.0
100-watt extended service	14.8
Fluorescent	
two 24-inch cool white (approx. 20 watts each)	50
two 48-inch cool white (approx. 40 watts each)	67
two 96-inch cool white (approx. 112 watts each)	73
High intensity Discharge	
400 watt phosphor-coated mercury	46
1,000 watt phosphor-coated mercury	55
400 watt metal halide	75
1,000 watt metal halide	85
400 watt high-pressure sodium	100
Theoretical Maxima	
All input energy reradiated as pure white light only	200
All input energy reradiated as monochromatic light at 555 nanometers wavelength (peak sensitivity)	680

Sources: National Bureau of Standards, *NBS Technical Note 789* (Washington: U.S. Department of Commerce, July, 1973) pp. 95–96; Illuminating Engineering Society, *Lighting and Application* (October, 1973); Westinghouse Electric Corporation, *Lighting Handbook* (Bloomfield, New Jersey, Westinghouse Electric Corporation, January, 1974) p. 3-1 to 3-35.

It is also interesting to note that within every specific type of lamp, without exception, higher output lamps are more energy efficient than lower output lamps. However, it is wrong to draw the obvious conclusion that higher output lamps result in more energy efficient lighting systems and are therefore to be preferred under vitually all circumstances. While that conclusion may be more or less valid for general background lighting and for situations where high levels of illumination are required at large distances from the light source (e.g., highway lighting), since received illumination decreases as the square of the distance from the light source, it is quite possible that using a few low output lamps very near to the specific task that requires high illumination (e.g., in desk lamps), combined with background lighting, may be far more energy efficient overall.

Other Comparative Factors

Unfortunately, the comparison of alternative artificial light sources is not simply a matter of relative efficacies. There are a number of other technical and economic factors that need to be considered, including: life expectancy, color rendering ability, maintenance of lumen output, and first cost. In general, incandescent lamps have the shortest lives, the best color rendering ability, among the lowest deterioration of lumen output over time, and the least initial cost. Fluorescent lamps typically live longer, have a somewhat lower color rendering ability, a similar secular deterioration of lumen output, and a considerably higher first cost. Finally, HID lamps, which are normally the most expensive, have fair to high life expectancy, among the lowest color rendering ability, and a fair to good maintenance of lumen output.

It should be clear, from even this brief description, that these light sources are not perfectly interchangeable. Rather the priorities attached to the specific lighting parameters relevant to the particular set of tasks to be performed will color the choice. Nevertheless, it is clear that HID lamps are generally to be preferred on grounds of energy efficiency where high levels of lighting are required and fluorescent lamps where lower lighting intensities are appropriate. Whenever the quality of light produced by these lamps is acceptable, they should be used.

Natural Lighting

Of course, whenever it is possible to use natural lighting to replace artificial lighting it is desirable to do so. Natural light approaches the theoretical definition of pure white light, containing nearly equal amounts of all the visible wavelengths during the peak of daylight. It is also a free, renewable, and as close to pollutionless energy source as is conceivable.

There are, however, several problems with the use of sunshine as a light source. Probably the most important is that it is variable over an enormous range of intensity and this variation in the lumen output emerging from the sky is not controllable. A second problem is that the solar radiation reaching the earth's surface contains a significant ultraviolet component which, under some conditions of exposure, may be damaging both to the eyes and the skin. In addition, there is the problem of solar heat, again via the greenhouse effect, as well as heat loss through window openings, both of which have previously been discussed.

Fortunately, all of these problems can either be solved or mitigated by appropriate design. Considering the latter problems first, the difficulties posed by both ultraviolet radiation and heat gain and loss through windows can be greatly mitigated by simply bringing natural lighting in indirectly and/or by using windows of appropriate filtering and insulating glass. For example, when lighting a large, high-ceiling interior space, a series of windows located at the top of the wall (in the clerestory style common to churches) will provide soft, indirect interior illumination throughout a good part of the space. Using insulating glass for these windows will further reduce the heat transfer problem. Yet another way to achieve indirect natural illumination is to fit windows with shading overhangs and then to increase the reflectivity of the ground (or ledge or balcony floor) immediately outside the window. Natural light will then not enter the window directly but will rather bounce off the interior ceiling and thus provide the desired interior lighting.

If indirect lighting is used, the natural variation in lighting intensity will not be as severe, because the higher extremes of illumination will be eliminated. Adjustable exterior or interior shading devices (e.g., venetian blinds) can not only achieve a similar effect without a preestablished indirect natural lighting design but can also be used to counter some of the variation in intensity of natural lighting whatever the initial design. They do require somewhat more attention and this is clearly a liability. But the considerable ability of the human eye to readily adjust to visual comfort under conditions of moderate fluctuations in footcandle levels makes continual adjustment unnecessary. We simply do not have to (and probably should not) eliminate all of this variation.

Nevertheless, the intensity of natural lighting will at times drop too low to provide sufficient illumination, even during the peak daylight hours. At such times supplemental artificial lighting will be required. However, even if natural lighting always required supplementation (which it does not), it would still be energy conservative to make maximum use of it. Every footcandle of artificial illumination that is replaced with a footcandle of natural light saves an amount of energy roughly three times as great as the electrical energy required to produce it.

Natural illumination puts some constraints on architectural design. Yet, both Frank Lloyd Wright and Le Corbusier have created buildings that use it extensively. For example, Le Corbusier's pilgrimage chapel of the Notre Dame du Haut at Ronchamp, France uses light blocks of different sizes to provide a pleasant level of natural interior illumination. If architectural philosophies as different as those of Wright and Le Corbusier can be reconciled with natural lighting, we need have no fear of inducing dull uniformity by recommending its use.

Lighting Systems

Design

The specific strategies for the design of energy conservative lighting systems are fairly straightforward. The system should be designed to provide a relatively subdued but adequate level of general illumination throughout the inhabited building space. Provision should be made for supplemental task lighting controllable by the occupants, either in the form of additional lamps within the same general fixtures but wired to separate switches or completely separated fixtures. Variable controls that allow the lighting to be turned up or down (i.e., dimmers) are also feasible, provided they are designed so as to reduce the input electricity and not to merely shunt and dissipate a portion of the energy. In cases where a particular task requires a high level of illumination, the supplemental lighting is nearly always best provided by a separate fixture located close to the task object (e.g., a desk lamp).

Natural lighting should be maximally integrated into the system. Light sources located near window areas should always be on different switches than those located farther away. Consideration should be given to using light-sensing switches to turn these lights on and off automatically, particularly in corridor lighting. Skylights should be utilized where possible.

Flexibility not only in the switching pattern but also in the positioning of fixutres is a very valuable asset. For this purpose furniture-mounted luminaires should be considered, along with wall- and ceiling-mounted extension lamps, and even freestanding movable light fixtures.

Of course, the choice of efficient light sources will do much to reduce energy consumption. Though more efficient lamps tend to have a higher initial cost, they frequently save so much energy that they are good financial investments and are likely to become even more so as energy prices rise. One analysis of the energy implications of replacing the current largely incandescent lighting scheme of a typical three-bedroom apartment with fluorescent general interior lighting showed that 300 watts of fluorescent lamps could provide at least as much illumination as 1,260 watts of incandescent lamps and could do so with a better lighting distribution.[21] Furthermore, a cost analysis of these alternate lighting systems for a 1,000 unit housing project in New York City revealed that the higher initial cost fluorescent system paid for itself in under 2 years and had less than half the discounted annual cost of the commonly used incandescent scheme, at a

10 percent interest rate.[22] Unfortunately, the fact that tenants typically bear the operating costs of the lighting system whereas building owners bear the initial costs has almost invariably militated against investment in energy efficient systems, even where such investment is handsomely justified on economic grounds.

The design of lighting fixutres can be as important as the proper choice of lamps. Since the light generated in a lamp tends to radiate in all directions, some of the light will not be shining into the desired spaces. Fixtures should be designed to reflect all such light efficiently in directions that will allow it to contribute to illumination needs. A properly designed fixture should not only direct light but also diffuse it to eliminate glare without cutting the level of illumination inordinately. Excessive shielding of fluorescent light tubes, for example, has been common practice despite the fact that fluorescent tubes distribute light fairly effectively even without additional diffusers. However, if additional diffusion is required, adding a plastic louver with a 45 degree cutoff will reduce effective lamp output by only 24 percent, while a translucent plastic cover will reduce it by 35 percent, and mounting the tube in a suspended semi-indirect container shielded with plastic will produce a 75 percent diminution in delivered light.[23]

Luminaires designed to recover some of the heat generated by light sources are not only energy efficient from a heating standpoint, as discussed in the preceding chapter, but also contribute to lighting efficiency. This is because the lumen output of a given lamp tends to vary with ambient temperature such that peak output may be achieved at a temperature below that which will result if heat is allowed to build up in an enclosed fixture. For fluorescent lamps, maximum lumen output occurs at between 70° and 80°F ambient still air temperature, depending on lamp design.[24] Light output from the same fluorescent tubes drawing the same current can be 10 to 20 percent higher at this peak than at temperatures on the order of 100°F.

Proper fixture design can improve light quality and thus increase apparent illumination without altering lamp output. It can also reduce dirt buildup, which causes a deterioration in illumination produced (known as *luminaire dirt depreciation*). Fixture design can clearly make a very important contribution to energy conservation in lighting systems. It is also a wise idea to provide an automatic shut off system for all lights in office and public buildings (with individual area manual override) to guard against the carelessness that often results from the fact that individual occupants do not generally bear the costs of lighting directly.

Operation

Lighting Maintenance. The light output of electric lamps falls over time because of dirt buildup as well as physical deterioration of the lamp. To the extent that maintenance can reduce this decline, it will conserve energy be reducing the need to overdesign lighting systems to compensate for the gradual drop in the level of illumination. The effectiveness of lighting maintenance is illustrated by a study of one fluorescent lighting system.[25] At the end of a 3-year period, illumination was found to be only 60 percent of initial footcandles when no maintenance was done; when luminaires were cleaned every 1½ years, the levels were maintained at 68 percent of initial footcandles; and when cleaning was done once a year, along with simultaneous relamping of one-third of the luminaires, illumination was maintained at fully 78 percent of initial levels.

Although more frequent cleaning is nearly always energy conservative, more frequent replacement of lamps tends to be less so because it requires increased energy expenditure in the manufacture and transportation of the lamps. In any case, the high cost of labor relative to energy has tended to keep the frequency of maintenance far below that which would result in the maximum conservation of energy. Rapidly rising energy prices should tend to reduce this divergence in the future.

Janitorial and Other Maintenance Scheduling. It has become commonplace to schedule janitorial and other maintenance services before or after normal business hours to the maximum extent possible in order to reduce interference with operations. But doing so tends to increase energy consumption, particularly by lighting systems, by extending the hours that their operation is required. For example, a survey of 197 office buildings reported in *Electrical Consultant* (September, 1975) showed that the annual hours of lighting-use demand exceeded the annual hours during which the buildings were open by an average of 13 percent.

Careful rescheduling of janitorial and other maintenance services to allow maximal performance during normal working hours (to the extent to which this is possible without significant disruption of primary activities) should result in reduced lighting demand. Where services must be performed before of after hours, they should be done by teams of maintenance people covering one section of the building at a time. This will allow lighting (and other building) systems to be shut down in all sectors but that being serviced at any given time. This, of course, presumes that the building has been designed to allow for this sort of piecemeal control.

Light Source Operation. By now we have all heard the familiar injunction to turn off unneeded lights, and though this is a very simple step, it does have considerable value as an energy conservation measure. However, the question inevitably arises as to how long a light must be unnecessary before it pays to shut the light off. That is, if a light will be needed again shortly, is it better to shut it off or leave it burning? There is both an energy and an economic answer to this question.

For an incandescent lamp both answers are the same, because the start-up power in-rush is very small and the number of starts does not significantly affect the life of the bulb. For example, once a 100 watt incandescent bulb has been burning for about three-quarters of a second it has used up as much electricity as was consumed during the start-up.[k] Therefore, operationally it is always energy conservative to shut off incandescent bulbs no matter how short the period in which their services are not required. Likewise, since the bulb life is not affected, money will also be saved by following this immediate shut off policy.

The situation is different for fluorescent and HID lamps because there is a greater initial requirement for energy. More importantly, the life of these bulbs is very much dependent on the number of starts. From an operating energy viewpoint, the period required for the continuous burning of the lamp to consume as much energy as the start-up is very brief. But since the shortening of lamp life implies increased manufacturing and transportation energy, the total energy break-even period is somewhat longer. On the other hand, because of the relatively high initial cost of the bulbs (including installation, where relevant), the economic break-even period is significantly longer still. Nevertheless, although the exact calculation depends on a number of economic factors, even the economic break-even period is probably less than 15 to 20 minutes in most areas and in many cases considerably less. For example, according to one calculation for a 40 watt two-lamp ballast fluorescent light with electricity costs at 2.5 cents per kilowatt-hour and $1 labor cost per lamp replacement, the break-even period is only about 6.5 minutes.[26]

At any rate, as a rought rule of thumb, it is probably energy conservative to shut off fluorescent and HID lamps when the interval of non-use exceeds a minute or two—and certainly when it exceeds 5 or 6 minutes. Similarly, it is probably cheaper to shut such lamps off than to leave them on when their services will be unnecessary for 15 minutes or more.

Summary and Conclusions

Lighting accounts for nearly 25 percent of the annual United States consumption of electricity. Lighting standards in the United States have grown very considerably, more than tripling in the last 25 years. There is reason to believe that this escalation of lighting requirements has resulted in present-day illumination levels that are far in excess of actual physiological and psychological needs.

The most widely used standards, those developed by the Illumination Engineering Society, are strongly based on the laboratory research of H. Richard Blackwell. These standards constitute a wholly unrealistic extrapolation of results determined for virtually unidimensional laboratory tasks to multidimensional real world tasks. Reduction of illumination standards to levels less than half the

[k] According to Con Edison's "Customer News" (November, 1973) the power in-rush for a 100 watt incandescent bulb is 0.0213 watt-hour. Such a bulb burning for one second consumes 0.0277 watt-hours of electricity. Hence, after 0.769 seconds the bulb has consumed an amount of operating energy equal to the initial in-rush.

current IES recommendations would not only save vast amounts of energy but also produce lighting more attuned to human needs.

We have seen that although visual acuity increases over a broad range as footcandle levels rise, the marginal increments in acuity have no significance for the efficiency of performance of visual tasks such as reading above about 50 fc. Furthermore, there are a number of parameters relevant to the performance of what we call visual tasks besides mere quantity of light—parameters such as the degree of patterning of the task object, size and contrast, degree of glare, spectral properties of the light, the amount of eye adaptation permitted, time factors, and nonvisual correlates of the task. By evaluating and designing tasks with all of these factors in mind, it should be possible to achieve high quality task performance without excessive illumination and thus with considerably energy savings.

Whatever the ultimate decision on lighting requirements, it is energy conservative to select light sources with a high ratio of lumens per watt. Despite their relative cheapness and high color rendering ability, incandescent light bulbs are highly inefficient light sources, converting only about 10 percent of input energy into light. Fluorescent lamps are several times as efficient and last far longer, although they are considerably more expensive initially and do not render color as truly. High-intensity discharge lamps are more efficient still but are also more expensive and even worse at rendering true color.

In the future, it is quite possible that improvements in the technology of HID lamps will drastically reduce these disadvantages and make them far more attractive. Certainly research efforts directed to this end would be a worthwhile investment in energy conserving technology. For the present, their use should be maximized wherever their disadvantages are not critical (e.g., street and other outdoor lighting). Fluorescent lights should be used in all other cases, except where color rendering ability is given considerable importance, in which case, incandescent bulbs should be used. For general, large area lighting it is always better to use higher output bulbs, no matter whether incandescent, fluorescent, or HID lamps are appropriate.

Lighting systems should be designed to maximize the use of natural lighting. They should provide for an adequate level of general illumination along with supplemental, occupant-controlled, specific-task lighting. Flexibility in the control as well as the positioning of lighting fixtures should be optimized. All lighting fixtures should be well-maintained, including both cleaning and relamping.

Janitorial and other maintenance should be scheduled during normal operating hours to the maximum extent possible in order to minimize demand on lighting (and other building) systems. All before- or after-hours work should be done in teams on a sector-by-sector basis.

Finally, lights should be conscientiously shut off when they are not needed. While fluorescent and HID bulbs probably should not be shut off if they will be needed again within a few minutes. incandescent bulbs should always be shut off no matter how brief the interval of non-use.

Architect Richard Stein has estimated that lighting energy conservation measures, particularly the adherence to more realistic lighting standards, could potentially reduce lighting loads in schools, commercial buildings, and institutions by more than 50 percent.[27] Furthemore, he has estimated a 25 percent energy saving in the manufacture of lighting equipment. Since Stein does not include improved switching arrangements or optimization of the use of natural lighting, it seems that his estimates are likely to be only a lower bound on the percentages of national savings achievable by conscientious application of the complex of lighting energy conservation measures recommended here.

Addendum to Part I
A Note on the Tradeoff between First Cost and Operating Cost

Frequently a tradeoff exists between the initial cost of physical capital (e.g., a machine, a building) and its subsequent operating cost. That is, it is often possible to lower the maintenance, energy, etc. costs that will have to be paid out year by year by initially purchasing better quality capital, which typically has a higher purchase price. Thus one trades off higher capital costs for lower future operating costs, or inversely.

A rational individual seeking to minimize costs would balance present costs against future costs in some way appropriate to the evaluation of the tradeoff. For example, one might calculate the rate of return implicit in spending a given amount more now in order to achieve a greater stream of savings over a period of years. By such a criterion as rate of return, one could analyze which among the array of available capital represented the best buy and then act accordingly.

We have seen that added insulation, double windows, airlock doors, more careful design and construction, replacement of incandescent lamps with fluorescent and/or replacement of fluorescent lamps with HID lamps, etc. can all reduce the consumption and, thus, the cost of energy. In some cases, reduction of the initial cost of other systems (e.g., due to the possibility of using smaller heating and cooling systems) more than offsets the higher initial costs implied by the energy conserving designs and procedures discussed. In most other cases, the added cost of the conservation measures would produce a fairly respectable to very high rate of return in terms of energy cost reductions. Why then have these procedures not typically been followed purely out of economic self-interest?

Let us first consider those cases in which the net effect of the energy conservation measures is to increase first cost. Since some of these measures would have paid off even at pre-1973 energy prices (though with a much lower rate of return), one would have expected them to be more widely instituted than they have been. Part of the reason for this lies in the institutional separation of builders and building users.

The builders knew that they would not own the buildings during their operating years. They further knew that a lower first cost was an important asset in selling buildings built speculatively (i. e., without a contracted buyer) as well as being vital to obtaining contracts awarded on the basis of lowest bid. Thus there was an incentive for builders to bias the tradeoff in the direction of lower initial and higher future costs. Their informational advantage relative to the buyers resulting from greater knowledge of building construction made it possible for them to sell such buildings without having to deal with critical evaluations of construction alternatives by buyers. On the other hand, the buyers' lack of sophistication and bias in favor of focusing more strongly on first costs pressed builders into avoiding higher initial cost designs with lowered operating costs. Where the operating energy costs were ultimately to be borne by building tenants, as is frequently the case with respect to electricity (at least for lighting), the buyers' bias in favor of lowered initial costs was strengthened. From several directions then the deck was stacked against energy conservation.

But what of the situations in which a savings even in overall initial costs was achievable by following an energy conservative approach to design? Why were not all these options fully exploited?

It must be assumed that part of the answer, at least, lies in the ignorance that derives from following established designs without ever considering whether they can be improved upon or, for that matter, whether or not they are appropriate to the particular geographic location and functional requirements of the building at hand. Designing with these kind of blinders on should always be avoided, no matter what our design criteria or the relative priorities they bear. To be sure, one cannot cope with designing every building or, for that matter, anything else completely from scratch. But surely, it is the obligation of any designer to always be on the lookout for ways to achieve at least marginal improvements in the performance of the design object. And periodically it does make sense to sit down and play with the possibilities for achieving the design objectives in fundamentally different ways. After all, is this not the creativity, the fun, in any design work?

Partially these sort of blinders arise because we have been taught that certain areas are not a fruitful point of focus for our attention. This was largely the case with energy in the United States. Energy was so cheap and so abundant for so long that little serious attention was given to the possibilities for achieving direct economic benefits from reducing the consumption of energy in most cases. Only when energy was clearly a dominant cost, e.g., in electricity generation, was it a focal point of design attention. Now as energy becomes not only more expensive but also more restricted and more a topic at the forefront of consciousness, it can be expected that at least the most economically attractive of the energy conservation alternatives will increasingly be exploited. Unfortunately though, old habits die hard and there is thus likely to be considerable inertia in this process without more direct incentives and publicity.

To some extent legal restrictions add to the inertia of this process. Design codes written in terms of specific prescriptive design requirements rather than performance requirements for the finished product tend to hamstring creativity and freeze us into existing technologies. For example, Caudill, Lawyer, and Bullock (*A Bucket of Oil,* Boston, 1974) cite a case in which their design for an energy conservative, compact cold weather schoolhouse ran afoul of a state code because it provided for skylights rather than windows in some of the classrooms. The design, which was well-received by the local school board, teachers, and students, was vetoed by the state, and they were forced to return to a traditional, energy-wasting design.

Thus the failure to exploit many even economically attractive energy conservation opportunities in the past was probably the result of a combination of lack of sophistication, unduly restricted thinking, lack of attention, and inordinately detailed prescriptive building codes. Some of these problems are already being dealt with, but all of them must be dealt with before we can expect to see the kind of fluidity of response to money-saving energy conservation opportunities that a simplistic view of economic theory would lead us to expect.

Part II
Transportation

5 Transportation and Energy

Introduction

The transportation of people and goods accounted for almost one-quarter of the total energy and more than one-half of the total petroleum consumed in the United States in 1970.[1] Nearly three-quarters of this amount was consumed in the movement of passengers, while the remainder moved freight.[2] This imbalance is at least partly due to the relative inefficiency of transporting people: goods can usually be packed more closely together, and they do not ordinarily require either ventilating systems or comfortable chairs.

In order to fully exploit the potential for conserving energy, it is necessary to: use an energy-efficient combination of transportation modes, coordinate the different means of transportation effectively, improve the energy efficiency of each mode as much as possible, and reduce the overall need for transportation. Using the proper combination is particularly important because there is such a wide variation in energy efficiency from mode to mode. Improved coordination, on the other hand, not only saves energy by reducing energy consuming delays (e.g., by aircraft forced into holding patterns) but can also increase the effectiveness of the entire transportation network. Redesigning and adjusting the operation of each individual means of transportation with a view to maximizing its energy efficiency will clearly conserve energy even in the absence of either a favorable shift in the transportation mix or improved coordination. Finally, if the amount of transportation required to support a given level of economic and social activity can be decreased (by means which, in themselves, are not overly energy intensive), further gains in energy conservation can be registered.

Energy Efficiency and the Intercity Modal Mix

Comparative Energy Efficiencies

The average energy efficiencies of each of the major modes of intercity passenger and freight traffic are presented in Table 5-1. The degree of intermodal variation is impressive. The most energy efficient freight mode (pipeline) is able to move a given weight of cargo more than 93 miles with the same energy

Table 5-1
Energy Efficiencies of Major Intercity Transportation Modes, 1970

	British Thermal Units/Passenger-Mile	
Passenger Modes	*Actual Load*	*100% Load*
Bus	1,600	740
Railroad	2,900	1,100
Automobile[a]	3,400	1,600
Airplane	8,400	4,100

Freight Modes	*British Thermal Units/Ton-Mile*
Pipeline	450
Railroad	670
Waterway	680
Truck	2,800
Airplane	42,000

[a]The energy efficiency of the automobile of the mid-1970s is undoubtedly lower due to the detuning of engines and the attachment of various fuel-wasting emission control devices.
Source: Adapted from Eric Hirst, "Transportation Energy Use and Conservation Potential," *Bulletin of the Atomic Scientists*, November, 1973, pp. 38–39.

that the least efficient mode (airplane) requires to move that cargo only a single mile. Even a comparison between two more similar modes of freight transportation, trains and trucks, reveals that the former is more than four times as efficient as the latter. Although the variation in energy efficiency for passenger traffic is not as great as for freight, it is nevertheless considerable. The most energy intensive mode (airplane) requires more than five times as much energy per passenger-mile as the least intensive mode (bus), whether evaluated at actual or potential load. Likewise, the automobile, the least energy efficient of all the land-based modes is more than twice as energy intensive as the passenger bus.

In the case of passenger transportation, the load factor achieved is as critical as the choice of mode. For example, even though bus travel was more than twice as energy efficient as automobile travel in 1970, a fully loaded automobile would have consumed the same energy per passenger-mile as a bus at average actual load. The energy-saving potential of increasing load factors highlights the value of improved scheduling, carpooling, etc. Still this does not downgrade the importance of choosing the proper transportation mix, because shifting traffic toward less energy intensive modes not only increases the fraction of traffic they carry but also improves their load factors. Thus a double saving is achieved. As long as the reduction in traffic carried by other modes has its primary effect in reducing the number of vehicle trips rather than in reducing their vehicle load factors, the full saving will be

Table 5-2
Percentage Distribution of Intercity Traffic by Major Transportation Modes

Mode	*1950*	*1955*	*1960*	*1965*	*1970*	*1974*[a]
Freight[b]						
Oil Pipelines	11.8%	15.7%	17.2%	18.6%	22.3%	22.8%
Watercraft	14.9	16.7	16.6	15.9	16.5	15.5
Railroads	57.4	50.4	44.7	43.7	39.8	38.0
Trucks	15.8	17.2	21.5	21.8	21.3	22.7
Aircraft	0.0	0.0	0.1	0.1	0.2	0.2
Passenger[b]						
Buses	5.2%	3.6%	2.5%	2.6%	2.1%	2.1%
Railroads	6.4	4.0	2.8	1.9	0.9	0.8
Automobiles	86.2	89.0	90.1	88.9	86.6	85.8
Airplanes	2.0	3.2	4.3	6.3	10.0	11.0

[a]Estimated from preliminary data.
[b]Modes listed in order of decreasing energy efficiency.
Sources: U.S. Bureau of the Census, *Statistical Abstract of the United States: 1974* (Washington: U.S. Department of Commerce, 1974), p. 547; American Association of Railroads, *1974 Book of Facts*, p. 636.

realized. But, for example, shifting passengers from air to rail will not produce this full double saving if airlines still schedule the same number of flights and run them one-third rather than one-half full. To a certain extent the economics of the situation will prevent this from happening; high fixed costs (fixed in the sense of being independent of the percentage of load achieved) per vehicle-trip will discourage operation at excessively low load factors.

It should therefore be obvious that even if nothing else were to change, shifts in the distribution of passenger and freight traffic among the existing means of transportation can easily produce very considerable reductions or increases in transportation energy requirements. This naturally leads to the question of how the flow of traffic is, in fact, distributed among the various modes and how the pattern of this distribution has been changing over time.

The Distribution of Intercity Traffic

The historical pattern of the distribution of intercity traffic among the major modes is presented in Table 5-2. For freight, the dominant feature has been a relative shift of traffic away from rail and into more energy efficient pipelines, on the one hand, and less energy efficient trucks on the other. This at first appears to be an energy conservative trend, since the pipelines gained nearly twice as much traffic (in percentage terms) as did trucks. But this trend has actually increased energy intensivity. For each ton-mile of cargo shifted from rail to pipelines 220 Btu's were saved, but for each ton-mile shifted from rail to trucks 2,130 more Btu's were consumed. Thus the trend tould have had no net energy effect only if the relative shift to pipelines had been nearly 10 times as great as the shift to trucks. Since the relative gain of pipelines was considerably less than that, a significant increase in the overall energy intensivity of freight transportation resulted. The secular growth of highly energy-inefficient air cargo has also aggravated this situation.

The passenger transportation scene has been dominated by the automobile. Least energy efficient of all the means of land transportation, it carried between 85 and 90 percent of all intercity travel. There has been a reversal during the last decade and a half of the increasing trend in the automobile's relative share of traffic. However, this reversal was achieved at the expense of an increase in air transportation, an even more inefficient mode, and thus was anything but energy conservative. In fact, air travel is the only mode that has gained relatively. The secular decline of the least energy intensive passenger transport modes, bus and train—particularly the latter—has been nothing short of spectacular. Thus the average energy intensivity of intercity passenger traffic has also undergone a considerable secular increase.

Determinants of the Traffic Distribution

Since we are interested not simply in conserving energy but in doing so with minimal negative effect on the standard of living, we must seek to understand what technical and economic factors produced the observed pattern of traffic distribution, and why. In particular, attention should be focused on those considerations relevant to the efficient functioning of the transportation system as a provider of transportation services.

Speed, flexibility, and cost have cleraly been primary factors in influencing the distribution of traffic. The premium placed on speed arises from the view of geographic distance as an impediment to social and economic activity. For example, the hundreds of miles intervening between the site of the production of a box of cereal in Michigan and the site of its consumption in, say, New York City enforce a delay between these two activities at least as long as the time required to transport the cereal between these two locations. The greater the speed of transportation the shorter the delay. Higher speed transport will, *ceteris paribus,* reduce spoilage, decrease the size of in-transit inventories, and increase the facility with which goods and services can be spot-ordered to cover unforeseen circumstances. Thus the more susceptible goods or people are to one or another form of deterioration, the higher their unit value, and the greater the degree of emergency involved, the greater emphasis there will be on speed.[a]

The flexibility factor includes both the ease with which routings can be altered and the degree to which cargo with varying characteristics and requirements can be accommodated. If routes may be readily altered, it will be easier to minimize door-to-door delivery time, particularly where points of origin and/or destination are variable. Greater route flexibility also contributes to reliability, allowing alternate routes to be chosen when initially specified routes become clogged because of breakdowns of preceding vehicles, poor weather conditions, strikes of terminal operations personnel, etc. Adaptability to alternative cargo facilitates high capacity utilization when there is not enough cargo of a single type to fully load the vehicle thus reducing cost per cargo unit-mile (as well as increasing energy efficiency).

The cost of transportation, like any other cost, is important to any firm operating in a market place and any consumer interested in achieving maximum value for his or her given income. But the full cost of travel is not measured by the direct transportation charges alone. The travel time required also has a cost associated with it, since time is a valuable resource. For passengers this time cost may be roughly approximated by the amout of income that a given individual is capable of earning during that period of time. For freight the time cost is partly associated with the income that could be earned by an alternative investment of the money tied-up, as it were, in in-transit inventory and partly associated with any direct losses that may occur because of the time delay (e.g., loss of a sale to an impatient customer). Losses incurred by virtue of damage to, or deterioration of, the cargo in transit must also be included.

It should be clear that the speed, flexibility, and cost criteria are interrelated. A faster mode of transportation will reduce the portion of cost associated with time delay. A more flexible mode may increase door-to-door speed, especially when unforeseen route blockages occur, while also reducing direct travel cost per unit to the extent that its flexibility allows fuller utilization of capacity.

Presumably the energy consumption consideration would be taken into account through its affect on cost. But, as we have seen before, the relatively low price of energy associated with its relative abundance prevented the energy component of transportation cost from historically assuming a major role. In

---

[a]A person may be subject to abnormally high deterioration for medical or psychological reasons. For example, someone who is bleeding internally and in need of hospitalization will be adversely affected by any extension of travel time. Similarly, the higher value-to-weight criterion also applies to individuals if by higher value we are speaking in purely financial terms. For example, a firm transporting a high-priced consultant will tend to be more concerned with the travel time delay than one transporting a salesperson.

most cases the costs associated with time delay, capital equipment, and labor were dominant. Since there was no real weight given to the direct criterion of energy minimization either, energy consumption assumed little importance in the choice among transportation modes available (or for that matter, in the process of design of those modes).

The Case of Freight. The relative shift of freight traffic away from railroads and into trucks was due partly to the greater flexibility of the latter and partly to differential government subsidization. The inflexibility of rail routes creates problems not only because it, for the most part, renders door-to-door delivery infeasible without supplemental means of transit but also because a breakdown of one train on one section of a rail route can easily clog the route producing extensive delays for following trains as well. Thus a multiplicative scheduling snarl can be caused by a single failure.

The fact that trains require dedicated roadbeds (i.e., roadbeds usable only by the railroad) whereas trucks share roadbeds with other public and private motor vehicles may have played some role in the comparative neglect of rail facilities in the doling out of governmental subsidies for the construction, maintenance, and improvement of transportation route facilities. In any case, the net effect of this differential subsidization has been to generate an increasing gap between the coverage and quality of the highway system and that of the rail network. Truckers have also had a cost advantage because they have not had to bear as large a fraction of the expense of building and maintaining their own roadbeds as have the railroads.

The growth of air cargo has clearly been due in part to its substantial speed advantage, particularly for long hauls. But air traffic has also benefitted very considerably from high levels of government support, particularly in the research and development of aircraft, and to a lesser extent in the construction of airports and airport servicing facilities.

Although pipelines are extremely inflexible, their share of traffic has expanded because improvements in pipeline technology have made them a low cost–high reliability means of transportation for the cargos they are able to carry. Their efficiency as transporters of fluid fuels has allowed them to grow as such fuels have grown in importance.

Passenger Traffic. The enormous routing and scheduling flexibility of the private automobile has given it tremendous appeal as a means of passenger transportation. Its comfort and speed have been roughly comparable, if not superior, to other available means of land transportation, and it has additionally afforded an unexcelled level of privacy. The same massive attention to the maintenance and improvement of the road system that has so greatly benefitted the trucking industry has also enhanced the perceived advantages of the automobile. The cheapness of fuel and the comparative productive efficiency of the automobile manufacturing industry also made the cost of car travel competitive with other means. It is therefore no surprise that the private car came to dominate the passenger transportation scene.

The decline of rail passenger traffic has been accentuated by an increasing deterioration in the reliability, safety, comfort, and even the speed of rail transit. As one illustration of the latter, in October 1975 Amtrack (the U.S. rail passenger transportation corporation) announced the restoration of New York–Chicago rail service discontinued in the early 1960s. The new train would travel over the same route as the famous *Twentieth Century Limited* once did but would require 24 hours to complete the journey formerly completed in 16 hours.[3] Rail passenger service suffers from all the same maladies as rail freight service, but to an even greater degree.

Air passenger service, on the other hand, has rapidly increased in speed, reliability, safety, and comfort. This, combined with falling costs and expanding schedules, has made air travel increasingly attractive, especially for long hauls. The speed advantage of air travel has essentially overcome its origin to destination inflexibility. It is also quite possible that when full costs of travel (direct costs plus time costs) are included, air travel may, for many individuals, be the cheapest mode as well. Its rapid relative growth is thus not surprising.

Altering the Intercity Modal Mix

The joint goals of energy conservation and pollution reduction would both be served by a shifting of intercity passenger traffic from automobiles and aircraft to trains and buses and freight traffic from trucks and air to trains and water transport. But how do we accomplish these shifts?

Transfer Subsidy. The first and most obvious suggestion is to at least remove the government subsidy bias that has conveyed artificial advantages to the more energy intensive modes, if not to reverse it entirely. There is nothing inherent in the nature of rail travel that renders it incapable of becoming a fast, reliable, efficient, and attractive means of transportation for both passengers and freight. The fine rail systems that operate in nations as diverse as land-locked, circular Switzerland and the linear island nation of Japan make this absolutely clear. But rail service is not so inherently advantageous that it can be expected to overcome the kind of "stacking of the deck" in favor of its competitors that has proceeded for decades in the U.S. In fact, at this point it is quite possible that nothing less than a massive, though temporary, infusion of capital and personnel into the railroad system will be capable of overcoming the now self-perpetuating deterioration that is the result of decades of neglect. At any rate, a switch of subsidization emphasis from airways and highways to railways certainly has the potential of producing considerable energy savings. Of course, while a switch of funding towards the rail system may be necessary, it is certainly not sufficient. To produce such gains in a socially efficient manner, the funding must be intelligently applied and the resources commanded by those funds carefully directed, or the railroads will not be rendered capable of sustained, efficient operation but will merely constitute another addition to the nation's industrial welfare rolls.

Improve Coordination of Efficient Modes. Bus travel is probably best developed as a well-coordinated supplement to intercity rail (and air) travel, rather than as a substitute.[b] Linking bus and rail terminals and schedules would greatly improve the overall origin-to-destination flexibility of rail service, reduce travel time, and increase energy efficiency. Improved coordination and interconnection of water and rail facilities could also enhance the effectiveness of both modes.

Establish a Better Perspective on Speed. As will be discussed subsequently in more detail, increasing the speed of transportation typically imposes a considerable penalty in terms of energy consumption and often in terms of added pollution as well. Thus it is worthwhile asking whether the gains associated with higher speed can reasonably be expected to exceed the full costs (direct plus environmental) of achieving it.

Whether passenger or freight cargo is being transported, it is important to note that the relevant speed is neither the top speed nor the cruising speed of any given mode of transportation but rather the average door-to-door speed (effective speed) for the entire trip. It is often the case, particularly on short hauls (up to a few hundred miles), that modal interconnection delays will drastically reduce the effective speed of a journey undertaken primarily (in terms of distance) by very high speed transit.

For example, the author recently traveled from Cambridge, Massachusetts to Yonkers, New York (both suburbs) by a combination of buses, subways, and airplane. The largest part of the approximately 225 mile distance, the journey between Logan Airport in Boston and La Guardia Airport in New York, required about 45 minutes, takeoff to landing. But, the door-to-door time was nearly five hours. Thus, though the average speed of the air travel was roughly 250 miles per hour, the effective trip speed was closer to 45 miles per hour. The trip would almost certainly have been faster by automobile—and perhaps nearly as fast by train, despite the low quality of passenger rail service between those two areas.

---

[b]Having once taken a six- to eight-hour trip on an intercity bus operated by a major bus line, the author (who is average sized for a male born in the United States) can personally attest to the discouraging discomfort that directly results from the lack of seat space, including leg room, on such prolonged trips. Subsequent trips between the same cities have always been made by car, chiefly for that reason.

The central point is a very simple one. To the extent that speed is important, it is not *modal* speed comparisons that are relevant but *effective trip* speed comparisons. Normally the apparent relative speed advantage of highly energy intensive air travel over far more energy efficient rail travel is considerably reduced (if not eliminated) for short to medium hauls when effective speed comparisons are made. This results from the fact that airports must, for reasons of safety and space requirements, be located reasonably far outside of the main business or residential areas of cities, whereas trains may come right into the heart of the city.

Furthermore, a simple-minded bias in favor of higher speed is not always in the best interests of the firm or individuals paying for (or engaging in) the transportation, even where the direct charges per ton-mile (or per passenger-mile) are equal. If, for example, a product that has been produced and readied for shipment will not actually be used at its point of destination until a specific, more or less known, future time, it may be cheaper to use a relatively slow means of transport and take advantage of the free storage that it provides rather than rushing the cargo to its destination and paying warehousing charges there. For passenger travel the equivalent may be the substitution of a longer, overnight train (or boat) ride for a faster plane ride, in order to save the added expense of one night at a hotel, while not losing any awake time. Of course, if the faster mode is also more expensive, as is usually the case, the argument for using the slower mode is that much stronger.

Reducing Travel Requirements

During the twenty-five year period beginning with 1950, the volume of intercity freight traffic increased by about 110 percent, to a 1974 total of more than 2.25 billion ton-miles. Passenger traffic rose even more rapidly, increasing more than 160 percent in the same period, with nearly 1.33 billion passenger-miles logged during 1974.[c,4] Hirst has estimated that rising per capita traffic was responsible for more than half the increase in energy consumed in intercity and urban passenger transportation and nearly two-thirds of the rise in intercity freight energy consumption during the 1950s and 1960s.[5] Thus transportation has increased considerably in the past decades, both in absolute and relative terms.

Because of the physics of the situation, as discussed subsequently, transportation of people and freight would inevitably require considerable energy

[c]To get some perspective on the huge amount of travel represented by these figures, an individual traveling at 1 mile per second (3,600 miles per hour) would require more than 42 years of continuous travel to equal the amount of intercity passenger travel for the single year 1974. A ton of freight moving at the same speed would have to travel for an entire human lifespan (72 years) to equal the 1974 intercity freight totals.

expenditure even if it were carried out efficiently. It would therefore make sense, from an energy conservation viewpoint, to explore the potential for reducing travel requirements (rather than treating the volume of traffic and its growth rate as a given), even in a world far closer to the ideals of energy efficiency than we are likely to come. The case for such a review is clearly that much stronger because of our inefficiency.

There are three major paths to the reduction of travel requirements: substitution, locational alterations, and life-style changes. In keeping with our consistent bias towards minimizing negative effects on the standard of living resulting from energy conservation, the first path—utilizing less energy intensive substitutes for travel—will be emphasized. Changes in the patterns of geographic location will also be considered within the context of maintaining or improving the standard of living. Life-style changes of a relatively minimal nature will be discussed, but larger scale alterations will be ignored, primarily because of the extremely subjective character of any evaluation of the effect of such changes on the standard of living.

Substituting for Travel

The prime energy effective substitute for transportation is telecommunications (i.e., communication at a distance). Under present technology this includes telephone, television, radio, teletype, and other facsimile transmission and reception. It is far from a perfect substitute for a number of reasons, not the least of which is that there are whole classes of transportation that cannot even theoretically be replaced by such equipment (at least given present and likely near future levels of technical knowledge).

Most freight transportation falls into this category, with the exception shipments of materials that are already merely written communication (letters, books, bills, periodicals, etc.). Transportation of production workers (as opposed to administrative employees) to and from jobs in the manufacturing, mining, construction, agricultural, and fishing industries is also not replaceable by communications, nor is transportation related to medical and dental services or on-site inspection or protection work (e.g., insurance claims adjusters, building inspectors, security guards). Nevertheless, communications may be capable of considerably reducing travel related to the provision of nonphysical personal services (e.g., consulting work, education) to the performance of managerial/administrative/clerical work and to the carrying out of various social and consumption related activities.

A 1974 study performed for the Urban Mass Transportation Administration attempted to estimate the potential reduction in urban passenger travel achievable by substituting telecommunications for more direct interpersonal interactions.[6] Their estimate, on the basis of admittedly optimistic assumptions of substitutability[d] (including availability of the full range of telecommunications equipment wherever required, ignoring costs), was that between 14 and 22 percent of all urban area vehicle miles traveled might be avoidable.[7] Applying these figures to the estimated 710 billion urban passenger-miles traveled in 1970[8] yields and estimated potential saving of between 99 and 156 billion passenger-miles for that year. Since more than 95 percent of urban passenger-miles were traveled by automobile in 1970,[9] assuming a generous average gas mileage of 15 miles per gallon for this urban driving implies that a maximum of 6 to 10 billion less gallons of gasoline would have been required for urban auto travel during that one year.

In any given instance, the substitutability of telecommunications for travel depends, of course, on the character of contact involved. Where the contact is predominantly unidirectional, as in the giving of instructions or directives, substitutability is high, and relatively simple communications modes are usable, e.g., telegraph. If interaction of a primitive nature, e.g., verification of receipt of instructions, is required these simple modes may still be adequate. But as the requirement for response and counter-response increases, it becomes far more efficient to utilize a communications mode designed for at least basic interactive communication—the telephone. A great variety of interactive contacts may be handled with reasonable efficiency by this mode. However, the telephone, as such, is subject to two very important limitations: it can only be used to transmit the verbal or, more properly, the aural component of human communications;[e] it is not capable of transmitting collateral materials (e.g., photographs, reports, contacts) which may be helpful or even critical to the effectiveness of the contact.

Contacts of a more sensitive nature (including, but not restricted to, contacts with an important emotional component) are greatly enhanced by the ability to perceive nonverbal signals. Since television or videophone modes of telecommunications provide for some limited transmission of these so-called nonverbals, they may be usable as a substitute for direct contact in some of these circumstances.[f] These modes also have considerable advantage insofar as they are able to transmit some collateral materials. However, the various sorts of

---

[d]The following assumptions were made with respect to the percentage of trips that were substitutable: 75 percent of trips related to clerical work; 65 percent related to professional and technical work; 50 percent related to family shopping; 25 percent to educational, civic, religious, and miscellaneous family business; 20 percent of the trips made by managers, officials, and proprietors; 5 percent of sales worker, and 5 percent of social and recreational trips.

[e]The importance of nonverbal communications between humans has been increasingly recognized in the past few decades, fostering the development of a field of study called *kinesics.* See, for example, R.L. Birdwhistell, *Introduction to Kinesics* (Louisville: University of Louisville Pess, 1952).

[f]Future technological developments, particularly in the field of holography (i.e., wavefront reconstruction photography), which allows the projection of truly three dimensional

facsimile transmission equipment are probably superior modes with respect to this latter criterion.

Of course the successive introduction and improvement of each given mode of communication has already provided some degree of substitution for direct contact and hence for travel. But since the improvement of telecommunications has not only made some substitution possible but has also made indirect (i.e., telecommunicative) contact easier and cheaper, it has undoubtedly stimulated a vast growth in the aggregate number and frequency of contacts that take place. Thus there is a problem in determining how far this substitution has already proceeded, which would be of value in estimating to what extent the potential for substitution remains unexploited.

One way of getting at the magnitude of the residual substitution potential is simply to survey individuals involved in activities in which the contact requirement is high. Such a study was conducted among British civil servants in the early 1970s by a group at University College, London.[10] Workers surveyed were asked what percentage of an aggregate total of 6,400 face-to-face contacts in which they had engaged could have been carried out equally well over telephone or telex. For the percentage of contacts that did not fall into this category they were asked why these contacts could not be satisfactorily handled by these modes.[g] Where the objections to telecommunications could be overcome by various higher capability modes (e.g., videophone, facsimile transmission equipment) it was assumed that these contacts too would be substitutable were these modes available. Though only 3 percent of these contacts were held to be substitutable by telephone or telex, almost two-thirds were estimated to be substitutable by one or another of the range of telecommunications equipment.[11]

A series of experiments in communications performed by the New Rural Society Project (NRSP) of Stamford, Connecticut casts light on the operational implications of the different capabilities of alternative communications systems,

---

images, may eventually decrease the disparity between direct contacts and telecommunications even further. This technology, which interfaces strongly with laser technology, will have to undergo a very considerable advancement before such capabilities as long distance transmission of holographic images are even on the horizon. The energy requirements of this future technology are thus largely unpredictable. For an interesting lay discussion of this fascinating field see Keith S. Pennington, "Advances in Holography," *Scientific American* (February, 1968); for a more recent and more technical discussion see David D. Dudley, *Holography: a Survey* (Washington: National Aeronautics and Space Administration, 1973).

[g]The experimental methodology employed here is at least one step removed from that which is directly relevant to the substitutability issue, since the workers involved did not have the more sophisticated modes of communication available. They were thus expressing opinions based on incomplete knowledge and experience. Their estimates of the primary reasons why telephone and telex were not substitutable, in the 97 percent of the cases where nonsubstitutability was alleged, may have been biased in either direction by this lack of experience (as well as other considerations). Therefore the assumption that the ability of a superior mode to overcome this primary objection (to the use of telephone or telex) meant that that mode was, in fact, substitutable may have been unwarranted.

as well as on their substitutability for face-to-face contacts.[12,13] One experiment involving audio-only telecommunications indicated that increasing the number of sound channels did not affect the acceptability of communications between pairs of individuals, while a second experiment found evidence that multichannel audio is a real advantage when groups of more than two people are involved. It is also interesting that in seven out of nine cases discussions over three-channel audio systems were judged as good as those using a video system, a much more expensive alternative.

In a pilot study, most of 20 business managers surveyed by NRSP said that 50 to 80 percent of their present face-to-face meetings involving three or more people and a minimum of 15 minutes of travel could be effectively handled by an audio-only system. A second study indicated a predominance of agreement that a visual channel is not essential to routine conferences with people who are well known but is necessary where new people are present. This latter finding is undoubtedly due to the importance of nonverbal signals in gaining a more complete picture of the personalities and viewpoints of newly encountered individuals.

Social contacts are likely to be, on the whole, far less substitutable than business contacts. Those who have observed the apparent affinity of some individuals or social groups for frequent and lengthy telephone conversations may doubt the validity of this statement. However, though there is little evidence on this point, it seems far more probable that such communications are supplementary, i.e., that they represent contacts that would either not have taken place in the absence of this communications capability or contacts that would have been far less frequent.[h] At any rate, it will be assumed that the potential for reducing social contact travel by telecommunications is quite limited.

Before pursuing such policies as the large-scale substitution of telecommunications for face-to-face contacts, great care must be taken to understand their potential social psychological effects. In particular, there is a real question as to whether or not the reduced direct human contact that would result from this substitution would produce an increased depersonalization of society. Plausible a priori arguments can be made in either direction. It is possible that the travel time saved will be used partly to extend the length of actual (though indirect) contacts and partly to augment time spent in contact with proximate individuals, thus augmenting personal contacts. However, it is also possible that increasing the degree to which people relate to each other as disembodied voices or images on the face of an electronic box will increase isolation and decrease the extent to which others are perceived as real, flesh-and-blood human beings.

Probably the greatest danger of increasing the depersonalization of society is the possibility that it will lead to an increase in the level of violence—particu-

---

[h]It is, after all, the frequency and not the extensivity of contacts that is relevant to the potential for energy saving through reduced travel.

larly violence committed at a distance. It is a well-established principle of military training that making an enemy seem less human makes it easier to commit violence against that enemy.[i] Combining depersonalization with the increased capability for committing violence at a distance (i.e., violence that is itself depersonalized) provided by modern weapons technology is exceedingly dangerous.

That substituting telecommunications for face-to-face contact could have significant social/psychological effects derives from the simple fact that no individual human being is an entirely self-contained unit. We all interact with and are affected by our physical and psychological environment. Therefore substantial changes in this environment can be expected to have some effect on our behavior, though the magnitude and direction of this effect may be subject to debate.

The central point is not that the large-scale substitution of telecommunications for direct face-to-face contacts *would* be harmful but rather that it *could* be, and that the probability of deleterious effect is not a priori insignificant. Furthermore, the potential danger, if the effects work in that direction, is considerable. Therefore we need to exercise great caution in evaluating the extent to which this apparently effective energy conservation policy should be employed.

Locational Changes

Although individuals may sometimes derive pleasure directly from the act of traveling, it seems unlikely that any significant percentage of overall passenger traffic is explicable in these terms. Likewise, though the fact that a good has been transported from a distant location sometimes carries a degree of charm in and of itself, it is doubtful that this will go far in explaining the magnitude of freight mileage. Primarily, both passenger and freight transportation may be looked upon as necessary evils, required to overcome the barrier of geographic distance in service of the attainment of some particular goal. Whether the goal is recreational, business, or social, people travel to reach a location at which that goal may be effectively achieved, and freight is moved because it is desired at a location different from that at which it is produced. Therefore it should be possible to reduce transportation by an appropriate alteration of the spacial relationships between locations at which the various business, consumption, and social activities can be efficiently carried out.

There are basically two locational strategies: centralization and decentralization. If it were possible to centralize all varieties of all activities at a single point, the need for travel would clearly be minimized. However, the physical im-

[i] For example, Nazi concentration camp personnel referred to the human beings they were murdering in such huge numbers as "pieces" not as people.

possibility of doing so opens up the possibility that travel requirements may be reduced more by a considerable dispersal of various activities than by a maximal concentration of each separate activity at its own specific location. That is, if it were possible to concentrate employment, shopping, recreational, cultural, and residential activities successfully within a geographically small area, such a total city would drastically reduce the need for passenger travel. But, if it is not possible to provide a high quality of all of these activities within a city, it seems that the centralization of employment and cultural activities in a central city surrounded by bedroom residential and recreational suburbs may well be more travel-inducing than a more homogenized, decentralized locational pattern. Since the pattern of partial or specialized centralization referred to has in fact become common throughout the U.S., the centralization/decentralization issue reduces to: should the focus be on moving employment and cultural opportunities out of the central cities or on moving residential and recreational opportunities into the central cities?

From the point of view of reducing travel requirements, either change would appear to be desirable. However, the energy efficiency advantage of fully utilized mass transit modes over the (typically poorly loaded) individual transportmodes commonly called for in smaller communities, combined with the existence of considerable economies of scale[j] in the construction and operation of mass transit facilities, implies that the greatest energy advantage probably lies with the latter strategy.

Though there is no paucity of analyses of traffic generation and distribution patterns, most urban planners have centered attention of the problem of eliminating traffic congestion as opposed to reducing aggregate travel. That these are different problems is illustrated by the fact that locational patterns that smooth out rush hour peaks or enlarge the geographic area over which traffic is distributed will tend to reduce congestion, though they may well leave the aggregate amount of travel unchanged—or even increase it. On the other hand, a reduction in travel will *ceteris paribus* also reduce traffic congestion, so the two objectives are not always mutually antagonistic.

The lack of previous recognition of the minimization of travel, in and of itself, as a major planning objective implies that the possibilities for minimizing travel through innovative locational schemes have not been thoroughly explored. Since this is an area with potentially significant payoff in terms of energy savings, it is deserving of serious analysis.

It should be emphasized that the idea of innovation relating to locational schemes includes the development of architectural designs that overcome the perceived disadvantages of energy efficient centralized locational patterns. For

[j]The term *economies of scale* refers to the reduction in the cost per unit of product (in this case transportation services) associated with large-scale production. Economies of scale exist because there are some technologies of production that are highly efficient only at high rates of output (e.g., assembly line techniques).

example, since one of the reasons for the flight of residences to the suburbs is the desire for more living space, including a front lawn and garden, if it were possible to design a high-rise structure that provided these amenities, people might be attracted back to the central city.[k]

The increased use of high-rise buildings within central cities may have considerable potential as an energy conservation strategy. Providing the buildings are well-designed and are not too tall, they are more energy efficient than low-rise buildings in the supply of building services. The on-site location of a so-called "modular integrated utility system" could further augment this efficiency, using the by-product heat of on-site electric power generation for space and water heating, air conditioning, and liquid waste treatment.[14] Furthermore, high-rises could be used to free more land surface within urban areas for generalized recreational purposes, thus making city life more pleasant. The physical possibility of such a scheme is illustrated by the fact that even New York City, noted as a city of skyscrapers, has an average building height of only about six stories.[15]

If the high-rise buildings are designed for multiple uses, they may reduce travel requirements even further. The John Hancock Center in Chicago is probably the best illustration of the extremes to which the idea of multipurpose, multistory structures may be carried. Within this single building are located apartments, offices, recreational, and shopping facilities.

It is difficult to quantitatively estimate the net savings in energy that could be achieved by significant alterations in locational patterns, though it is clear they are considerable. By one estimate, derived through the use of mathematical models for simulation, arranging land use so that housing for a wide range of income groups and high-density employment areas are closely tied together could produce a 15 to 20 percent reduction in local automobile travel.[16] If it were in fact possible to achieve such a reduction in urban passenger traffic across all urban areas, that would represent a saving of between 100 and 135 billion passenger-miles, or 3 to 4 billion gallons of gasoline annually. However, given its rather abstract character, this cannot be directly taken as a serious estimate, though it probably is indicative of the order of magnitude of potential savings.

At any rate, it should be understood that locational change for the purpose of minimizing travel energy requirements is a rather long-run policy—much more so than the substitution of telecommunications for travel. It is also somewhat more difficult to achieve, since it requires the rearrangement of considerable quantities of fixed capital, particularly structural and infrastructural capital both of which are expensive and long-lived. In addition, the set of economic incentives (e.g., the possibility of increasing profits or sales) which would encourage private business to undertake the provision of extensive communications

---

[k] One way this might conceivably be achieved is through architectural designs that would provide enlarged terraces arranged alternately from floor to floor in the manner of leaf arrangements that maximize the incidence of natural light on the terraces.

facilities is not as strong—if it exists at all—in the case of locational alterations. Finally, altering locational patterns is far more politically disruptive and may be far more socially disruptive than many of the other energy conserving strategies suggested.[1]

Life-Style Alterations

The discussion in the previous section focused most strongly on the locational problem as it relates to passenger as opposed to freight transportation for two reasons: first, passenger transportation is by far the largest energy consumer; and second, the reduction of freight travel requirements is more a matter of life-style considerations than a matter of optimizing clearly suboptimal locational patterns.

The United States is a physically immense nation, within whose boundaries there is enormous geographical variation in climate, topography, and resource availability. Yet we have become increasingly accustomed to utilizing cheap transportation (fueled by cheap energy) to homogenize the availability of product on a year-round basis throughout the nation. The most striking illustration of this is in the sphere of agricultural products that are often shipped thousands of miles in order to provide various markets with produce that would otherwise be out of season in that area. As long as people in New York are willing to pay the price for lettuce shipped from California whenever New Jersey lettuce is out of season, the lettuce will be shipped cross-country at the expense of considerable energy consumption.

The necessity for retarding spoilage of agricultural produce transported long distances to satisfy the demand for year-round availability typically implies the use of an energy intensive preservation technique (e.g., refrigeration), which increases transportation energy requirements even more. Furthermore, the trend toward increasing purchase of convenience foods, especially those which are precooked then frozen, leads to still higher energy-use in transportation (as well as in processing, retailing, and home storage), because the foods must be kept frozen and not just refrigerated. Thus a turning away from convenience foods as well as a reacceptance of seasonality would surely reduce freight transportation energy requirements.

There is a rather strong interface between the locational and life-style issues. The development of land-use patterns, in general, and their residential component, in particular, are clearly influenced by the population's concept of

---

[1]The political disruption would result from the redistribution of population and business tax bases that would benefit certain local political entities at the expense of others, as well as from possibly extensive rezoning requirements. The social disruption would primarily be a function of population relocation.

preferred life style (within the constraints of economic feasibility). The striving of families in the U.S. to attain the suburban, home-owning, two-car-family life style, which was an integral part of the post–World War II concept of the "American Dream," resulted in a massive expansion of resident population in the areas surrounding central cities. This frequently resulted in the transference of land from agricultural to residential uses. As the amount of agricultural production in close proximity to the cities declined, it became more and more necessary to transport agricultural products to the central cities and their suburbs from increasingly distant places. In some cases, the land removed from agricultural use was quite fertile and productive. It was this process which, for example, produced the progressive elimination of truck farms on Long Island and resulted in the necessity of transporting potatoes from Maine and Idaho rather than Long Island to the New York metropolitan area markets. Of course, this same pattern of suburban sprawl also increased passenger transportation requirements as the ranks of the suburban commuter grew.

A changed concept of the good life, which prescribed high-rise living in the centers of cultural and commercial activity, might well produce a net reduction in both freight and passenger travel over current patterns, provided these revitalized central cities utilize available land intelligently and provided some of the land in surrounding areas is reclaimed for agricultural purposes. But this involves a massive change in life style—a change that though potentially energy conservative may not be the optimum way of achieving the goal of reduced energy consumption while maintaining or improving the material standard of living.

Some Aspects of The Physics of Transportation

Newton's Laws of Motion

To begin with, it is necessary to differentiate clearly between the concepts of velocity and acceleration. *Velocity* refers to the speed at which an object moves in a given direction, while *acceleration* refers to the rate at which velocity changes. For example, an object moving along a straight path fast enough to travel 10 miles in 1 hour is said to have a velocity of 10 miles per hour (mph) in that direction. If, rather than moving at a constant speed of 10 mph, the object is caused to increase its speed by 20 mph during every hour that it travels, so that at the end of 1 hour its speed is 30 mph (rather than 10 mph) and at the end of 2 hours its speed is 50 mph, then the object is said to be accelerating at the rate of 20 mph per hour. Although strictly speaking acceleration refers to any change in velocity, frequently it is used to refer to an increase, while deceleration is used to refer to a decrease.

Isaac Newton, the principle architect of the field of physics, now known as classical mechanics,–the field relevant to our transportation energy concerns–summarized his system in the form of three laws of motion, the first two of which are most important here. Newton's first law, sometimes known as the *law of inertia,* may be stated as follows: Every body that is at rest tends to stay at rest and every body that is in motion tends to stay in motion at a *constant velocity* (along a straight line) unless it is acted upon by net forces. A *net force* is a force that is not balanced by an exactly equal force acting in the opposite direction. Thus when all forces *are* exactly balanced it is said that all net forces equal zero.

The first part of this law holds that if we wish to move an object originally at rest, we must push or pull it–and that is certainly congruent with everyday experience. But the second part of the law seems to imply that once we have exerted a force and caused an object to move, we can remove the force and the object will still continue to move at the same speed along a straight line. Continued force must be exerted only to speed it up, slow it down, or change its direction. Why then is it necessary to continue to expend energy, say by burning fuel in the engine of a vehicle, to provide continual motive force in order to maintain a constant velocity?

The answer lies in the fact that in the real world the forces that we exert or cause to be exerted on an object in order to move it are only a fraction of the totality of forces that act on that object. The forces that the world exerts on the object may oppose its motion, as do frictional forces and fluid-dynamic drag (the resistance of air and water to the motion of objects passing through), or may tend to make it move in directions that we consider undesirable for our transportation purpose, as does gravity in the case of an airplane. Since it is the whole group of forces acting on an object that determine whether and how it moves, in order to cause the object to continue to move at the same speed in the same direction, we must expend energy to create forces which balance or cancel out the effects of the perverse (from our transportation viewpoint) forces tending to decelerate or misdirect the object. In other words, as long as we wish only to have the object continue to move at a constant speed in a straight line, we need only insure that there are no *net* forces acting on the object, so that it can do what it would naturally tend to do in the absence of any forces.

Newton's second law can be stated: a given *net* force applied to an object will cause it to accelerate at a rate equal to the magnitude of the force divided by the mass of the object. In other words, a constant net force applied to an object will cause it to accelerate continuously and not just maintain its velocity. Furthermore, the amount of force required to achieve a given acceleration is larger the greater the mass of the object involved.

Work, Power, and Energy

When a force applied to an object causes it to move over a given distance, the force is said to have done *work* on the object. The amount of work done is equal to the product of the amount of force applied and the length of the distance over which movement was produced. If, for example, a thrust of 110 pounds causes an object to move 5 feet, the thrust force is said to have done 550 foot-pounds of work. The concept of work does not include any consideration of the time involved. That is, whether the 110 pound thrust had taken one second or one year to move the object 5 feet, the amount of work done would still be 550 foot-pounds.

Power, on the other hand, does involve a time consideration. It is defined as the rate at which the work is done. Thus power is greater if either more work is done in a given amount of time or less time is required to do a given amount of work. For example, a machine that performed the work described above in one second would be delivering 550 foot-pounds per second, defined as one horsepower. If two seconds were required, the machine would be delivering 0.5 horsepower. It is appropriate to describe the capacity (or output) of an engine in units of power, such as horsepower or watts, because that indicates the rate at which the engine is capable of (or currently is) performing mechanical work. Furthermore, if the power output of an engine is known, along with the amount of time it has been operating, the work performed by the engine during that period may be found by calculating the product of the power and the time. Therefore work may also be expressed in units of power-time, such as kilowatt-hours, as well as force-distance (such as foot-pounds).

There are two types of mechanical energy that an object may have: kinetic energy and potential energy. *Kinetic energy* is the energy associated with the motion of the object; *potential energy* with its position. For example, the energy of a bowling ball rolling down a bowling lane is kinetic energy. The energy of a bowling ball held still at a height of say four feet from the floor is potential energy—but if the ball is dropped, its potential energy will be transformed into kinetic (motion) energy, as the force of gravity causes it to fall to the floor. The kinetic energy K of an object is given by $K = \frac{1}{2}MV^2$, where M is the object's mass and V is its velocity. Thus it depends more strongly on velocity than it does on mass. The potential energy of an object, which is most important to the transportation problem, is that associated with gravity. If we choose the surface of the earth as our reference point and define potential energy to be zero there, the potential energy P of an object is given by $P = Mgy$, where y is the vertical distance of the object above the earth's surface, M is its mass, and g is a constant (related to gravity).

According to Newton's second law, when a net force acts on an object, it causes the object to accelerate. But since acceleration means a change in velocity and since kinetic energy depends on velocity, the application of the force will result in a change in the kinetic energy of the object. If the net force applied becomes zero, the kinetic energy will cease to change and the object will continue to move at constant velocity as per Newton's first law. If the application of the net force carried the object further above the earth's surface, its potential energy will have also changed. Thus, when a force is applied to move an object over a distance, i.e., to do work, it will result in a change in either the kinetic or potential energy of the object or both. Therefore doing mechanical work on the object is fully equivalent to transferring energy to it (or from it).

The Energy Required for Transportation

Consider the problem of transporting an object, initially at rest, from one place to another. First we must apply sufficient force to the object to overcome its inertia (i.e., tendency to stay at rest, in accordance with the first law) and any friction between the object and the surface on which it rests. We must apply enough force in the direction in which we want it to move to overbalance all other forces acting on the object. This net force will accelerate the object (in accordance with the second law) increasing its kinetic energy. If the force is directed so as to raise the object above the earth's surface, its potential energy will also be increased. When the object has reached the desired velocity (and height), the force we are applying must be reduced and then maintained so that the net force on the object is zero. Its kinetic and potential energy will thus remain unchanged. When the object approaches its destination, the kinetic and potential energy we have imparted to it must be removed. As the force we are applying is reduced further, the natural forces that resist the object's motion (e.g., friction, aerodynamic drag, and gravitational force) will come to dominate and a net force will result that tends to decelerate the object (and cause it to move closer to the earth's surface), until it is finally brought to rest at its destination.

The key question with which we are concerned is, of course: What are the critical factors that determine the amount of work *we* must do, i.e., the amount of energy *we* must transfer to an object to transport it? Let us first analyze this with respect to the simplified single origin–single destination situation just described.

Since the object is initially at rest ($V = 0$), the amount of kinetic energy that must be transferred to the object is $K = \frac{1}{2}MV^{*2}$, where V^* is the object's maximum velocity. It is therefore clear that the amount of energy required for transportation depends on both the mass of the object and the speed at which it is transported, but that *increasing speed incurs a larger energy penalty than increasing mass.* Though it will take twice as much energy to transport an object whose mass is twice as large at a given speed, it will take four times as much energy to transport an object of a given mass twice as fast. Thus, apart from any considerations of engine efficiency, it is inherent in the physics of the situation that, for example, making cars lighter and reducing effective speed limits will reduce the transportation energy requirements associated with automobiles.

Frictional and fluid-dynamic drag forces operating on the object do negative work, i.e., they tend to remove energy from the object. Thus the amount of energy we must transfer to the object to allow it to maintain a given level of kinetic energy depends on how strong these forces are. The force associated with friction depends most strongly on the weight of the object and the properties (chiefly texture) of the contact surfaces of both the object and the medium on which it rests.[m] Heavier objects again tend to require more energy, as do objects with rougher surfaces riding on rougher surfaces (e.g., rubber tires on an asphalt road as opposed to smooth steel wheels on a smooth steel train track). Fluid-dynamic drag forces tend to increase with increases in the speed with which the object moves through the fluid. The resistance of a fluid to the passage of a solid object also depends on the shape and position of the object relative to the direction of movement in the liquid. Thus the amount of energy we must continuously supply to the object to make up for its tendency to lose kinetic energy due to resistive forces tends to be greater the heavier the object, the higher the speed at which it moves, and the worse its fluid-dynamic profile.

If the transportation situation at hand also requires an increase in the potential energy of the object, we must also supply energy for this purpose. The amount of energy we must supply is given by $P = Mgy^*$, where y^* is the maximum height to which the object must be lifted above the surface of the earth. Since g is approximately constant for the problems with which we are concerned,[n] the energy required is greater the larger the mass of the object (the product Mg is actually the weight) and the greater the height to which it must be raised. Because velocity plays no role here, the mass of an object that is to be lifted above the earth as well as transported to a different surface location plays a somewhat greater role relative to velocity in determining transportation energy requirements than does the mass of an object traveling by surface transport only.

Combining all these considerations, it is clear that the total energy which we need to supply in order to transport an object from a point of origin to

---

[m]The so-called coefficient of kinetic friction also tends to fall as speed increases, implying higher speeds yield lower frictional forces and thus less energy required to overcome friction. But this effect is rather slight for speed changes in the ranges relevant to transportation. For example, the coefficient of kinetic friction for unlubricated steel on steel falls about 50 percent when speed is increased by a factor of 10,000. See Robert Resnick and David Halliday, *Physics for Students in Engineering and Science, Part One* (New York: Wiley and Sons, 1960), p. 98).

[n]In actuality, g decreases with altitude but does so very slowly. For example, its value at about 50,000 feet is only 0.5 percent less than at the earth's surface.

a single destination (both on the earth's surface) tends to be greater the larger its mass (or weight), the faster it is to be moved, and the greater the height to which it is to be lifted. In general, transportation energy requirements tend to be more sensitive to speed increases (especially for a given mode of transit) than to mass increases, though this tends to be somewhat less true for objects transported by air.

Until now we have been considering the simple case of a single acceleration to a desired velocity, maintenance of that velocity, and then deceleration to a stop. However, nearly all real-world transportation is considerably more stop-and-go than that, and this introduces an additional consideration into the energy requirements picture.

Every time an object being transported must be decelerated and reaccelerated, its kinetic energy must first be reduced then increased. If the removal of kinetic energy from the object needed to slow it down is accomplished by dissipation of that energy, then we must supply "new" energy to speed it up again. If, instead, the kinetic energy removed from the object is stored, then reacceleration may be accomplished by taking that same energy out of storage and re-using it. Thus the energy that we are required to supply to achieve a given amount of transportation can be greatly reduced.

It must be clearly understood that any process that transfers energy out of a moving object will slow the object down. There is no special magic attached to the use of frictional braking, in which energy is removed, for example, from the wheels of an automobile by forcing the spinning wheels to rub against non-spinning brake shoes. Such a braking process dissipates the kinetic energy as heat (thus contributing to the rate of increase of entropy) and is accordingly very wasteful. If instead a regenerative braking process were used, in which, for example, the kinetic energy of the spinning wheel was converted into electricity that was then stored in a battery, at least a portion of that energy (after generation, transmission, and storage losses) would be available for reacceleration.[o] The widespread use of frictional as opposed to some scheme of regenerative braking must therefore be seen as increasing the amount of energy that we must supply to carry out any given amount of transportation—an increase in transportation energy consumption that has no offsetting benefits and is clearly not required by the physics of the situation.

Collisions

The momentum of an object is defined as the product of its mass and its velocity (i.e., $p = MV$, where p is the momentum and M and V are mass and velocity). Thus the concept of momentum is related to, though different

[o]A regenerative braking scheme that seems to have considerably more promise than this one involves the use of flywheels. It is discussed in the following chapter.

from, the concept of kinetic energy (which is given by $K = \frac{1}{2}MV^2$). It is a basic principle of physics that when two or more bodies collide, the total momentum of all the colliding bodies is unchanged by the collision.[p] This is called the *principle of conservation of momentum.*

There are several distinct stages into which any collision may be divided. Consider, for example, the simple case of the head-on collision of two spherical bodies. First, there is the initial physical contact between the two bodies. Then there is a period of deformation during which the impact of the collision causes an alteration in the shape of the spheres—a squashing effect, as it were. This is followed by a period of restitution during which the forces generated by the resistance of the material of the spheres to the impact forces (which tend to squash it) cause the spheres to spring back to their former shape. Finally, there is a period of separation or rebounding.

If the period of restitution completely undoes the effects of the period of deformation, i.e., if the spheres spring back completely to their initial shape, the collision is said to be *elastic.* In such elastic collisions, the total kinetic energy of all the bodies involved is also unchanged by the collision. If instead the bodies do not completely regain their initial shape, we say that some permanent *plastic deformation* has taken place, and we call the collision *inelastic.* If the two bodies stick together after the collision, the collision is said to be *plastic.* In any case, in all inelastic collisions some kinetic energy is transformed into other types of energy, so that the total kinetic energy of the system is less after the collision.[q]

The amount of plastic deformation that takes place during an inelastic collision is clearly of great importance as far as the safety of the occupants or freight cargo of colliding vehicles is concerned. It can be measured by a coefficient of restitution, which equals one for a perfectly elastic collision (i.e., no plastic deformation) and represents larger deformation the smaller the value. It is normally a positive fraction, since virtually all real-world collisions are to some extent inelastic.

The value of this coefficient depends on the physical properties of the colliding bodies as well as their velocities. It will, of course, be far larger for a collision between two billiard balls, for example, than for a collision between two balls of clay. But for nonhomogeneous objects, such as automobiles, the determination of this coefficient is not simple.

It is probably easiest to understand which factors are most critical to the

---

[p] Strictly speaking this is only true in the absence of all forces (such as gravity or friction) external to the collision. But in practice, these forces are generally so small relative to the impulsive forces occuring during a collision that they can be safely ignored and momentum conservation assumed.

[q] Total kinetic energy does not necessarily become zero, even in the case of a completely inelastic collision. In fact, the total kinetic energy after the collision may also be greater than before, if some other form of energy is changed into kinetic energy by the collision.

effects of a collision on the objects involved if we first consider the simple case of a perfectly elastic head-on collision between two spherical bodies initially moving (without spinning) in exactly opposite directions. In such a collision, we know that total momentum is unchanged (true of all collisions) and that total kinetic energy is unchanged (because it is elastic). Thus, by equating the total momentum before the collision to the total momentum after the collision and similarly for kinetic energy, we have enough information to solve for the after-collision velocities of both the bodies, if we are given information as to their masses and their velocities before collision. This will enable us to calculate the kinetic energy of each before and after the collision and hence determine how energy is transferred during the collision.[r]

Making the simplifying assumption that the collision speeds of the two bodies are equal (though their directions are opposite) yields the following results: if the objects are equal in mass, they will rebound off each other with unchanged speed (although reversed direction) and there will be no transfer of kinetic energy from one to the other; if one object has a mass that is larger than the other, the larger mass will lose kinetic energy to the smaller mass.

During an inelastic collision we know that a part of the kinetic energy is converted into other forms. One way this kinetic energy is lost is by its dissipation in the process of deforming the objects involved in the collision. Since we know that in an elastic collision between unequal masses, the smaller mass gains energy from the larger mass (i.e., the share of the smaller mass in the total system energy increases), it seems logical that the smaller mass would have more energy to dissipate in the deformation process as a result of an inelastic collision. Thus, for objects made of materials with similar deformation characteristics, the smaller mass would be more severely deformed by the collision.

Therefore, in general, it may be assumed that in a collision between two objects of roughly similar structural properties but different masses, the smaller mass object will be more deformed. Likewise, higher initial velocities for any two given masses imply higher initial kinetic energy, thus requiring more energy dissipation during inelastic collision and, hence, *ceteris paribus* producing greater inelastic (or permanent) deformation.

Summary and Conclusions

Because there is considerable variation in energy efficiency among the different modes of transportation, changes in the distribution of traffic have substantial energy impact. Historically, there has been a relative shift toward the more energy intensive modes. In freight, the share of traffic carried by the relatively

[r]See the appendix to this chapter

energy efficient railroads has declined sharply, while that carried by much less efficient truck transportation had climbed. In passenger travel, the energy inefficient automobile came reasonably close to monopolizing intercity traffic, only to lose some ground in recent years to the still less efficient airplane. Passenger rail service, on the other hand, has all but disappeared from the intercity scene.

These changes have occurred partly because of some real advantages of the modes that have grown, partly because of the cheapness and abundance of fuel, and partly because of the precipitous deterioration of the alternative modes aided and abetted by direct and indirect governmental biases in favor of their more energy intensive competitors. It is clear that the share of traffic carried by the more energy efficient modes can be greatly increased, with consequent large energy savings. But it is also clear that in order to accomplish this reversal in the long-term trend without sacrifice in the standard of living, large intelligently applied infusions of capital will be required to make up for decades of neglect. However, this investment should produce a substantial triple return, conserving energy, reducing pollution, and improving the quality of transportation services.

In addition to shifting the distribution of traffic toward less energy intensive modes, it should be possible to substantially reduce transportation energy by simply reducing the overall volume of travel. For some large and important classes of transportation this may be accomplished by substituting telecommunications for transportation. Experiments have demonstrated the technical feasibility of at least some degree of substitution, and the potential energy savings from widespread adoption of this idea are quite large (6 to 10 billion gallons of gasoline per year). Nevertheless, it is important to explore the potential social and psychological effects of the large-scale replacement of direct human contact with telecommunicative contact before any such proposal can be seriously recommended.

Changes in the patterns of location of residential, business, recreational, social, cultural, and educational opportunities relative to each other could be made in such a way as to substantially reduce the need for travel. It would seem that the locational pattern with the most promise for simultaneously conserving energy and maintaining the standard of living is closer to the centralized city pattern than that of what has been called suburban sprawl. The subject of the minimization of energy consumption through the alteration of locational patterns is one that deserves and requires further study—study that may conceivably overturn the tentative conclusion stated above. But if that conclusion is supported, the energy conservation argument will be added to the various other social, political, economic, and even moral arguments for the revitalization of the nation's decaying central cities.

Life-style changes can play a role in making energy conservative locational changes feasible. They can also conserve transportation energy to the extent that the seasonality of foods is reaccepted and the demand for excessively processed foods reduced.

A brief review of some basic physical principles reveals that there is an inherent energy penalty (i.e., additional energy expenditure) associated with greater weight and higher speed. In general, the energy required to transport an object is more sensitive to increases in the latter than to increases in the former. Frictional and fluid-drag forces oppose motion and thus the larger they are, the more energy that must be transferred to an object to transport it.

If the transportation involves considerable slowing down and reaccelerating, energy must be alternately removed from and then transferred back to the object. If the energy is removed by dissipation, e.g., as in frictional braking, then additional "new" energy must be introduced for reacceleration. If instead the energy removed is stored, it may then be re-used when reacceleration is required. This regenerative braking procedure thus minimizes the energy we must inject into the system to accomplish the desired transportation.

In general, a collision between two bodies of similar structure will result in more damage to the object with lower mass. However, if two bodies of the same mass and structure collide, they will both experience a similar degree of damage, regardless of whether they are both large or small. In addition, the higher the speeds at which they collide, the more damage will typically result. Hence, the basic physics of the situation does not support the notion that small vehicles are inherently less safe during collisions with other vehicles, but it does imply that reduced vehicular speeds should contribute to increased safety.

Appendix 5A

Transfer of Kinetic Energy in an Elastic Collision

Let M_1 and M_2 be the masses of two spherical bodies. Assume that these bodies collide head-on, i.e., that they are initially moving with velocities U_1 and U_2, respectively, in exactly opposite directions. One of these directions will be taken to be positive and the other negative, so that $U_1 \geqslant 0$ and $U_2 \leqslant 0$. Assume also that the collision is perfectly elastic (i.e., that there is no plastic deformation). Let V_1 and V_2 be the postcollision velocities of masses M_1 and M_2, respectively.

Now the principle of conservation of momentum implies:

$$M_1 U_1 + M_2 U_2 = M_1 V_1 + M_2 V_2 \tag{5A.1}$$

And since the collision is elastic, kinetic energy is also conserved, thus

$$\tfrac{1}{2} M_1 U_1^2 + \tfrac{1}{2} M_2 U_2^2 = \tfrac{1}{2} M_1 V_1^2 + \tfrac{1}{2} M_2 V_2^2 \tag{5A.2}$$

The conservation of momentum equation (5A.1) can be rewritten as:

$$M_1(U_1 - V_1) = M_2(V_2 - U_2) \tag{5A.3}$$

And the conservation of kinetic energy equation (5A.2) can be rewritten as:

$$M_1(U_1^2 - V_1^2) = M_2(V_2^2 - U_2^2) \tag{5A.4}$$

Dividing (5A.4) by (5A.3) yields:

$$(U_1^2 - V_1^2)/(U_1 - V_1) = (V_2^2 - U_2^2)/(V_2 - U_2)$$

$$\Rightarrow \quad (U_1 - V_1)(U_1 + V_1)/(U_1 - V_1) = (V_2 - U_2)(V_2 + U_2)/(V_2 - U_2)$$

Therefore

$$V_2 = U_1 + V_1 - U_2 \tag{5A.5}$$

Substituting (5A.5) into (5A.3) we can solve for V_1 and then for V_2, the post-collision velocities:

$$V_1 = \left(\frac{M_1 - M_2}{M_1 + M_2}\right)(U_1) + \left(\frac{2M_2}{M_1 + M_2}\right) U_2 \tag{5A.6}$$

$$V_2 = \left(\frac{2M_1}{M_1 + M_2}\right)(U_1) + \left(\frac{M_2 - M_1}{M_1 + M_2}\right) U_2 \tag{5A.7}$$

In order to highlight the affects of mass inequalities, consider the special case in which the masses are initially traveling at the same speed ($U_1 = -U_2$).

Now, let $M_1 = aM_2$, where a is constant and $a \geqslant 1$.

Substituting for M_1 and U_1 in equation (5A.6) yields:

$$V_1 = \left(\frac{aM_2 - M_2}{aM_2 + M_2}\right)(-U_2) + \left(\frac{2M_2}{aM_2 + M_2}\right)(U_2)$$

$$\Rightarrow V_1 = \left(\frac{1-a}{1+a} + \frac{2}{1+a}\right) U_2$$

$$\Rightarrow V_1 = \left(\frac{3-a}{1+a}\right) U_2 \tag{5A.8}$$

Similarly, substituting in equation (5A.7) yields:

$$V_2 = \left(\frac{2aM_2}{aM_2 + M_2}\right)(-U_2) + \left(\frac{M_2 - aM_2}{aM_2 + M_2}\right) U_2$$

$$\Rightarrow V_2 = \left(\frac{-2a}{1+a} + \frac{1-a}{1+a}\right) U_2$$

$$\Rightarrow V_2 = \left(\frac{1-3a}{1+a}\right) U_2 \tag{5A.9}$$

Dividing equation (5A.8) by equation (5A.9) produces:

$$V_1/V_2 = (3-a)/(1-3a) \tag{5A.10}^*$$

Case 1: $a = 1 \Rightarrow M_1 = M_2 \Rightarrow V_1/V_2 = -1$

Therefore, when masses are equal, the final velocities are equal (but opposite). Also, $V_1 = U_2$ and $V_2 = U_1$, so that the masses only exchange velocities. But since their precollision velocities were equal, neither has experienced a change in the magnitude of its initial velocity. Thus their kinetic energies remain unchanged by the collision.

Case 2: $1 < a < 3 \Rightarrow M_2 < M_1 < 3M_2 \Rightarrow -1 < V_1/V_2 < 0$.

Therefore, their final velocities will be opposite in direction and $|V_1|$

$< |V_2|$. Since $U_1 = -U_2$ and total kinetic energy is unchanged, M_1 slows down and M_2 speeds up as a result of the collision. Therefore M_1 loses kinetic energy and M_2 gains kinetic energy, so there has been a transference of energy from the larger to the smaller mass.

Case 3: $a = 3 \Rightarrow M_1 = 3M_2 \Rightarrow V_1/V_2 = 0$.

Thus $V_1 = 0$. From equation (5A.9), $V_2 = -2U_2$. Therefore mass M_1 loses all of its kinetic energy to mass M_2.

Case 4: $a > 3 \Rightarrow M_1 > 3M_2 \Rightarrow 0 < V_1/V_2 < 1$.

Therefore, the postcollision velocities will be in the same direction (in the original direction of mass M_1) but $V_1 < V_2$. Since $U_1 = -U_2$ and total kinetic energy is known to be conserved, mass M_1 will be slowed down by the collision and mass M_2 will be speeded up. Thus there will still be a transference of kinetic energy from M_1 to M_2.

Note that $\lim_{a \Rightarrow \infty} V_1/V_2 = 1/3$, so that regardless of how large mass M_1 is relative to M_2 the transference of energy is always from the larger to the smaller mass, and the relative gain of kinetic energy by the smaller mass is never greater than when the larger mass is precisely three times as large as the smaller.

Of course, as the initial velocities are larger and larger, though the same relative amount of kinetic energy is transferred for any given relative mass sizes, the absolute amount of kinetic energy transferred grows.

6 Land Transportation: Automobiles and Trucks

In 1974 land-based transportation carried almost 85 percent of all intercity freight traffic, nearly 90 percent of all intercity passenger traffic, and virtually all urban traffic in the United States.[1] Data from a few years earlier (1970) indicates that a single land-based transportation mode, the automobile, was responsible for some 87 percent of intercity and 97 percent of urban passenger traffic.[2] The truck, the closest freight-carrying counterpart to the automobile, accounted for about one-fifth of intercity and essentially all of urban ton-miles of freight transportation.[3]

Since nearly one-quarter of total energy and more than half of all petroleum consumed in the U.S. were consumed in the transportation sector, and since that sector is so dominated by land-based modes, it is clear that even small per mile improvements in their energy efficiency would generate substantial energy savings. Furthermore, the large relative share of automobiles and trucks indicates that increasing the attractiveness of the more efficient modes enough to induce a substitution toward them would also result in substantial energy savings, without diminishing the standard of living. Therefore a dual strategy is clearly appropriate. On the one hand, attempts should be made to improve the energy efficiency of all modes, but with particular attention directed to the popular, though least efficient, automobile and truck. On the other hand, efforts should be directed to increasing the relative comfort, speed, and reliability of the most efficient modes.

This chapter focuses on only one arm of that strategy—the improvement of the energy efficiency of automobiles and trucks.

Power Plants

The near universal acceptance of the standard gasoline-powered, piston-equipped internal combustion engine as *the* power plant for automobiles and small trucks and the diesel engine as *the* power plant for larger vehicles might lead one to believe that technical or economic necessity had dictated this state of affairs. Nothing could be further from the truth. The selection and specification of these engines from among the plethora of designs available in the early years of motor vehicle development proceeded in accordance with a specific set of social and economic criteria. These may have been legitimate at the time but are increasingly inappropriate today. In particular, the cheapness of petroleum derived fuels and the apparent lack of necessity to attend to the environmental consequences of vehicle design led to extremely low priority being assigned to the criteria of energy efficiency and pollution limitation. This is clearly no longer acceptable.

Once the piston internal combustion engine and the diesel were essentially selected, nearly all research and development attention, such as it was, was focused on the refinement of these engines to the virtual exclusion of radically different engine designs, at least insofar as the major U.S. automakers were concerned. The time is past due to look seriously, once more, at the full range of engine designs for this class of vehicles. Despite the relative poverty of research and development efforts devoted to these alternative designs, some interesting progress has already been made. A number of different engine designs are evaluated in the following section. More detailed descriptions of their operation can be found in the appendix to this chapter.

Evaluation and Comparison of Alternative Engines

All the engines we are going to discuss may be classified as either internal combustion engines (ICE) or external combustion engines (ECE). In principle, external combustion engines have an important advantage in that the combustion process may be freely manipulated without worrying about the effects of these manipulations on the functioning of the power plant itself. Thus the design of the combustion chamber, the air-fuel mix, the temperature and pressure at which the burning proceeds, and even the type of fuel used can be selected in such a way as to insure maximally efficient combustion, i.e., maximum extraction of fuel energy and minimum generation of pollutants. Furthermore, the continuous combustion of the ECE designs is inherently easier to control than the intermittent, explosive combustion shared by all of the ICEs except the gas turbine.

On the other hand, continuous combustion subjects the material of the combustion chamber to sustained high temperature rather than intermittent short-lived bursts of high temperature. In general, structural materials are able to tolerate far greater temperatures for very short pulses than they are able to tolerate when continuously applied. Since combustion tends to be more efficient at higher temperatures (at least within the ranges relevant here), this provides a partial offset to the advantage of ECEs.

The Otto ICE. In order to meet air pollution standards, the inherently dirty Otto ICEs commonly used as standard automobile engines were detuned and fitted with various pollution control devices, "pasted on" to the same basic engine design. The immediate and persistent effect of these patchwork solutions was to drastically reduce the energy efficiency of an already inefficient engine design. In addition, engines so modified became afflicted with such other annoying and potentially dangerous ills as excessive stalling and hesitation, thus reducing one of the main advantages of this power plant—its driveability (i.e., quick, reliable response and high performance). The installation of catalytic converters, which allowed some retuning of engines and hence some mild improvement of fuel economy, does not appear to be a viable solution because of its mild effect, the possibility of its enhancing the emission of new pollutants, its general unreliability, and its resource-depleting nature.

Since the greatest barrier to the motion of an automobile is its inertia when at rest, an engine must deliver high torque (twisting force) to the driveshaft and hence to the drive wheels when the car is stationary in order to initiate motion. But, because the Otto ICE delivers zero torque at zero shaft speed, the only way this can be achieved with an Otto ICE is to decouple the engine from the wheels via a clutch and idle the engine at 500 to 600 revolutions per minute (rpm). This wastes fuel, generates additional pollution, and creates noise, so that an Otto ICE is a particularly poor engine for stop-and-go urban driving.

This engine is also relatively noisy and not as adaptable to variations in the type of fuel used as might be desired.

In short, continued reliance on and committtment to the Otto ICE is poor policy both from an energy conservation and from a pollution control viewpoint. If, however, continued reliance on the Otto ICE is postulated, attention should be directed much more to basic redesign (e.g., stratified charge engine) than to add-on devices.

The Wankel ICE. The elegant simplicity of the Wankel ICE makes it a very appealing engine and one that should accordingly be producible at low cost as compared to the Otto ICE. Because power output can readily be increased by increasing the number of rotor and chamber units with little weight increase relative to power increase, the Wankel potentially should be able to deliver the same power as an Otto ICE with considerably less weight and size. The unidirectional rotation of the Wankel (as opposed to reciprocal action) implies low noise and virtually no vibration. The fact that the inflow of gas-air and the outflow of exhaust are continuous (there are no valves) eliminates surging problems as well as valve malfunctions. The engine is also apparently adaptable to a considerable variety of fuels.

However, the problem of insuring tight contact between the points of the rapidly rotating triangular rotor and the stationary chamber wall has been persistent. If these contacts are not tight enough, the three chambers are not sufficiently sealed off from each other and gas leaks between them, severely diminishing the efficiency of the engine. The explosive nature of the combustion must also be counted as a real disadvantage in terms of achievement of complete combustion.

Current production versions of the Wankel automotive engine have experienced serious fuel economy difficulties and do not constitute a significant improvement in pollution emissions over the standard Otto ICE.[4] Yet the basic simplicity of design of the engine, its potential for a high power to weight ratio, and its lack of vibration all indicate that further attempts to refine the engine may have a real payoff, at least for certain applications. But the inherent difficulties with ICEs do not make it the prime candidate for an ideal automotive engine, either on grounds of fuel economy or pollution reduction.

The Diesel ICE. The greatest advantage of the diesel engine is its relatively high fuel economy; roughly 50 percent better than the standard pre–emission-controlled 1968 Otto ICE.[5] However, although the diesel is capable of meeting 1975 standards of the Environmental Protection Agency (EPA) with respect to carbon monoxide, unburned hydrocarbons, and nitrogen oxide emissions, it has smoke, particulates, and odor emission problems.[6] It is also a noisy engine, and high-powered versions are heavy and large.

As an automobile engine, the diesel has relatively poor performance in the sense that diesel-equipped vehicles accelerate slowly. This is partly because of the necessity of using low-powered diesels in automobiles (because of the excessive size and weight of high-powered versions) and partly because diesels operate efficiently over a relatively small range of rpm. Poor acceleration is especially dangerous in highway driving—particularly in entering a high-speed lane of traffic from a low-speed lane or highway entrance. Like the Otto ICE, the diesel delivers poor torque at low rpm's and hence needs to idle when the vehicle is at a standstill, wasting fuel and generating additional noise and air pollution, which makes it less than ideal for city stop-and-go driving as well.

Thus, despite the considerable energy efficiency of the diesel, its combined performance and pollution difficulties do not qualify it as a highly desirable automotive engine.

The Gas Turbine Engine. The continuous burning feature of the gas turbine reduces polluting emissions and makes possible the design of a high-powered engine about the same size and up to 25 percent lighter than the standard Otto ICE. However, this continuous burning also implies the necessity of limiting combustion gas temperatures to prevent the failure of turbine blades. For example, as reported in 1969 the maximum continuous temperature tolerated by practicable existing blade materials was on the order of 2,000°F, whereas instantaneous maximum temperatures reached in standard Otto ICEs were on the order of 5,000°F.[7]

Reduced combustion temperatures imply reduced energy efficiency. On the other hand, the fact that the primary mechanical motion in the gas turbine is rotary eliminates efficiency losses associated with the conversion of reciprocal motion into the required rotation. On balance, the current generation of gas turbines is no more energy efficient at full load than pre-emission-controlled 1968 Otto ICEs and is only about half as efficient at part load. It has been estimated that advanced versions (using ceramic materials capable of withstanding very high temperatures) might achieve as high as 30 percent better fuel economy than the 1968 piston engine at full load but would still be substantially less efficient at part load.[8]

The gas turbine also has relatively low noise, little vibration (due to the rotary motion), considerable adaptability to alternative fuels, and high performance. It has, however, been a fairly expensive engine to produce, because of difficulties with certain parts—particularly the turbine blades.

Despite its favorable pollution and performance features, the relative energy inefficiency of the gas turbine (especially at part load), along with its production expense, does not make commitment to general-use automotive versions of this type of power plant overly attractive.

The Stirling ECE. By virtue of being an external combustion engine the Stirling engine combines important fuel economy and pollution reduction advantages. The more complete combustion achieved in present-day Stirling engines results in energy efficiency equivalent to that of the diesel (i.e., 50 percent better fuel economy than pre–emission-controlled 1968 Otto ICEs), while simultaneously easily meeting the stringent 1976 EPA emission standards.[9] It has been estimated that advanced versions of this engine will achieve energy efficiencies up to 25 percent better than present versions without any sacrifice in pollution reduction.[10]

The lack of explosive combustion and the elimination of valves makes for quiet, low vibration engine operation. The Stirling operates well at partial load and has a wide range of speeds and favorable torque characteristics, simplifying the transmission system and providing for relatively efficient stop-and-go operation.[11] It also performs well, accelerating rapidly and responding quickly to changes in load. Oil consumption is nearly zero, because the combustion products are not in contact with moving parts, and the engine is long-lived and reliable.[12]

Although the Stirling ECE is smaller than or comparable in size to an Otto ICE, it is up to 25 percent heavier and requires a radiator 2½ to 3 times as large. Heater design and fluid sealing are problems. Perhaps the greatest disadvantage of the Stirling is its relative sensitivity to the kind of abuse and neglect to which it might be subjected by an uninformed and uninterested operating public.

On the whole, the Stirling is a very energy efficient, low-polluting engine with good performance characteristics. It is extremely desirable power plant for general automotive use.

The Rankine ECE. Though the Rankine engine possesses the inherent low pollution advantages of an external combustion engine, the size of present versions may be up to 50 percent greater and their weight up to 75 percent greater than a 50 percent more powerful pre-emission-controlled 1968 Otto ICE.[13] The fuel economy of Rankine-powered vehicles has thus far been a persistent problem. Buses fitted with large steam turbine versions of this engine have reportedly achieved energy efficiencies comparable to diesel buses;[14] but the fuel economy of standard automobiles powered by the Rankine ECE has been on the order of up to 50 percent worse than that achieved with the pre-controlled Otto ICE.[15] Though the fuel economy of some automotive Rankine ECEs has been reported comparable to 1976 pollution-controlled Otto ICEs (a fairer comparison), this must be considered intolerably energy inefficient for standard automotive use.

Rankine engines are not sensitive to climatic variations or changes in altitude. They deliver high torque at zero shaft speed and thus do not have to idle when standing still, making them quieter, still less polluting, and relatively more attractive from an energy efficiency standpoint in frequent stop-and-go driving. They have good performance characteristics, are not overly sensitive to less than optimum servicing, and run quietly. However, their estimated production cost could be up to three times as great as comparable Otto ICEs.[16]

In sum, the Rankine ECE may already be a highly attractive engine for large vehicles—particularly those with a heavy component of stop-and-go driving, e.g., urban buses. However, it cannot be considered a prime candidate for general automotive use in its present state of development. But the potential of the Rankine is such that additional development, directed to the improvement of fuel economy and reduction of production cost, might well yield a handsome return.

The Electric ECE. The fact that the combustion associated with electric engines takes place at a centralized power generating facilities has advantages both in pollution control and fuel economy. However, these benefits cannot be fully realized without the availability of an efficient energy storage device.

The present generation of electric vehicles suffers from relatively poor performance, accelerating slowly to top speeds of only about 40 mph.[17] Their range between battery recharges is severely limited (about 50 miles), and battery recharge time is excessive.[18] But these are not failings of the engine per se—they are due to reliance on power packs of inefficient standard lead-acid batteries. The engines themselves are highly reliable, long-lived (up to 20–25 years) with minimal regular servicing,[19] and achieve overall fuel economy only slightly worse than well-tuned diesels (about 12 percent less), despite the inefficiency of central plant power generation and losses associated with the batteries.[20]

Under present circumstances, electric engines are unacceptable for general automotive use, though they are clearly superior to Otto ICEs for powering medium to small urban delivery vehicles (e.g., mail trucks) and for other low range, low performance purposes. The future potential of electric vehicles may well be enormous, but its successful exploitation hinges critically on the development of superior energy-storage devices.

Optimality in the Design of Motor Vehicle Power Plants

In evaluating the various types of alternative engines, energy efficiency and low pollution generation have been considered the criteria of principle importance. However, the relative emphasis it is appropriate to place on other evaluative criteria depends strongly on the type of function that the vehicle incorporating the engine is expected to carry out. For example, rapid acceleration, high top speed, and long range are of little importance in an urban bus or taxi, while quietness and efficiency under stop-and-go and part load conditions are highly desirable.

Because different classes of vehicles have such different performance requirements, it seems a priori unreasonable to expect a single type of engine to be universally optimal.[a] On the basis of the evaluations presented in the previous section, it would appear that the Stirling engine has the greatest promise for general automotive use, the Rankine for large vehicles, and the electric engine for small urban and commuter cars. There is still considerable testing and refinement to be done, however, before such statements can be made with great firmness and confidence, and it is not yet clear whether pure or hybrid forms of these engines will prove to be superior. But one thing does seem clear, it makes no sense, either on grounds of energy or ecology, to persist in directing huge amounts of resources to the development and production of paste-on devices and minor design changes in service of the salvation of the Otto cycle internal combustion engine. There are alternatives with far more promise available, and neither the time nor money required to bring them into efficient production and operation is likely to be excessive.

Although the time required for fruition is probably only on the order of 5 to 10 years at maximum for any one of these engines, there may still be some value in considering less radical hybrid versions of the presently well-established engines as interim solutions to the joint problems of fuel economy and pollution control. For example, one interesting hybrid described in the 1973 Senate hearings on automotive research and development combines a standard small Otto or Wankel ICE with an electric engine connected to standard lead-acid batteries.[21] The electric engine is used as a booster to provide bursts of power when needed for acceleration and recharges from the ICE when boosting is

---

[a]Even a nearly complete dependence on two classes of engines, as is presently the case (the Otto ICE for cars and small trucks and the diesel for larger vehicles), may be unduly restrictive. On the other hand, it is highly unlikely that an explosion of engine varieties would be either technically or economically desirable.

not required. Because it allows the use of a very small ICE relative to vehicle size, and because the engine braking involves putting energy back into the battery, the hybrid is fairly efficient and does not require excessive pollution control devices to meet EPA standards.

Fuel Economy and the Design and Operation of Motor Vehicles

Since motor vehicles are responsible for such a large part of the annual U.S. energy consumption, it is worthwhile to ask what sorts of energy savings may be achieved by changes in their design and/or in the ways in which they are commonly operated. To this end, we consider the handful of basic parameters that are most critical to the energy efficiency of such vehicles.

Weight

In order to initiate motion in a vehicle at rest or to speed up a vehicle already in motion energy must be transferred to it. The amount of (kinetic) energy that must be imparted to achieve any given increase in velocity is larger the greater the mass, and hence the weight, of the vehicle. Similarly, more energy is required to overcome the motion-retarding frictional force between a vehicle and the ground, *ceteris paribus,* the heavier the vehicle. More energy must also be expended to move a heavier vehicle up any given hill.

Since it is necessary to transfer more energy to a heavier vehicle to start it in motion, it is correspondingly necessary to remove more energy in order to stop it or slow it down. Though it is true that normal frictional forces (which always oppose motion) are somewhat greater for heavier vehicles, this advantage in decelerating vehicles with greater weight is swamped by the necessity to remove far greater (kinetic) energy in order to overcome the inertia of motion. Furthermore, the greater the weight of a vehicle, the more energy required to oppose the accelerating effects of gravity and maintain velocity on any downgrade.

For all of these reasons, weight is one of the most important factors in the determination of the energy required to operate a motor vehicle. One estimate indicates that fuel economy in automobiles driven under city conditions decreases by 0.25 miles per gallon (mpg) for each additional 100 pounds of vehicle weight,[22] but comparisons of fuel efficiencies from 1974 EPA tests indicate the possibility that the relationship may be even stronger than this. For example, the American Motors Gremlin, the Volvo 144, and the Chevrolet Vega, all of which weighed roughly 3,000 pounds, achieved about 16 to 17 mpg (despite engine sizes ranging from 101 to 232 cubic inches displacement); while the Mercury Cougar, the Chevrolet Malibu, and the Oldsmobile Cutlass, all of which weighed approximately 4,500 pounds, were rated at 9 to 10 mpg (and had engines of the same 350 cubic inches displacement).[23] The implied tradeoff between weight and fuel economy from this comparison is roughly 0.40 mpg per 100 pounds of body weight (ignoring all other factors). Put differently, vehicles weighing 33 percent less achieved between 60 percent and 90 percent better mileage in these tests.

All of the vehicles mentioned above were powered by Otto ICEs. The relationship between weight and energy efficiency may well differ for other types of engines. Large diesels, for instance, are relatively more efficient, so that a diesel bus weighing seven times as much as a diesel taxi consumes only about five times as much fuel per mile.[24] Nevertheless, all other things being equal, higher weight *always* implies higher energy consumption.

Since the gross weight of the total cargo of a passenger automobile is usually under 300 pounds (less than two adult passengers) and rarely over 600 pounds (four adult passengers), it is clearly absurd from an energy standpoint to use 5,000 pounds of automobile to transport that cargo when 2,500 pounds will do as well. Even if we postulate the necessity of building luxury automobiles, there is no need to build vastly oversized and overweight vehicles. The 1972 Mercedes 250 and BMW 3.0 limousines are certainly as luxurious, prestigious, etc. as the 1972 Cadillac de Ville, for example, and yet their curb weight is nearly 1,750 pounds less.[25] These limousines, at approximately 3,150 pounds, fall into the so-called compact class of automobiles by the common U.S. weight-oriented definition!

The weight of the average U.S. automobile in 1973 was 4,275 pounds.[26] Without any other alterations in either the design or operation of the automobile, a reduction in automotive weight sufficient to bring this average down to 2,500 pounds would reduce the amount of fuel consumed in operating automobiles by 30 to 40 percent.[b] Thus, this single change alone would reduce total U.S. energy consumption for all uses combined by 4.0 to 5.5 percent and total U.S. petroleum consumption by 8.5 to 11.5 percent, with no effect on the standard of living. Considerable additional energy savings would also accrue as a result of the reduction in energy requirements for the manufacture of smaller, lighter cars.[c]

A reduction in weight of the magnitude indicated could be achieved without any redesign by simply altering the mix of automobiles on the roads in the direction of increasing the relative numbers of the so-called subcompacts. Three principle objections have been vociferously raised to this highly effective energy conservation strategy—a demand objection, a supply objection, and a safety objection. On the demand side, the U.S. consumer, it has been

[b] Assuming no increase in miles driven because of better mileage.

[c] The production energy savings would likely amount to only about 10 percent of the operating energy savings, though this is a very large absolute amount of energy.

argued, is so committed to the large automobile that he or she will never accept smaller cars without extreme duress. Yet in the face of rising gasoline prices, sales of small cars have risen dramatically, as evidenced, for example, by the jump in the share of imports in U.S. auto sales from 16.6 to 21.1 percent between the fourth quarter of 1974 and the first quarter of 1975.[d,27] In fact, even before the fuel prices began to rise rapidly, the share of imports in U.S. car sales underwent a long-term spectacular rise, nearly quadrupling between 1965 and early 1975.[e] Of course, sales of small domestic cars have also skyrocketed since 1973.

On the supply side, the conversion to small car production was said to be greatly hampered by the inflexibility of the U.S. automobile industry and alleged to require enormous expense and considerable time. Yet in December 1973 the "big three" U.S. automakers, under pressure of the rush to small cars, estimated the total cost of emergency conversion at $500 million. This one-time conversion expense could have been completely financed out of their 1973 profits. Furthermore, Ford Motors estimated that it would have achieved conversion of 50 percent of its manufacturing capacity to small car production by the next spring.[28]

As to the allegation that smaller, lighter-weight cars are inherently less safe, we have seen that it is not the absolute mass of a body that is relevant to the damage that body will sustain in a collision so much as the mass (and hence weight) of the body relative to the mass of the object with which it collides. A collision between a 5,000 pound car and a 2,000 pound car will tend, all other things being equal, to result in much more severe damage being sustained by the smaller car. But, if both cars are the same weight, whether that weight is 5,000 pounds or 2,000 pounds, they will both tend to sustain equivalent damage. Thus, as far as collisions between small vehicles and other vehicles are concerned, it is only in the period of transition between large average cars and small average cars that there is any potential safety disadvantage associated with smaller cars.

However, small cars tend to afford a better view of the road (e.g., the hood is less obtrusive), have shorter turning radii, and are in general more responsive and easier to control in conditions of emergency swerving. Because of these advantages and also directly because of their small size, it may well be easier to avoid a collision when driving a small car. Furthermore, the fact that a small car must absorb more energy in a collision with a large car does not in itself mean that the occupants of the small car must suffer more damage. It should be possible to design vehicles capable of absorbing substantial amounts

---

[d] Imported automobiles are generally smaller and lighter than domestically produced United States cars, and their resultant greater fuel economy is apparently the reason for their sales boom at the same time United States automakers' sales were declining.

[e] The market share of imports in 1965 was only 5.6 percent. (*Statistical Abstract of the United States,* 1974, p. 560).

of energy in a collision while maintaining the structural integrity of the passenger cab, within the constraint of a 2,000 to 2,500 pound overall vehicle weight. Thus it is not necessarily true that small cars involve serious net safety reductions even in the transition period. They certainly would not after the move to small cars has been achieved.

In sum, there is no reason to expect either consumer resistance, severe manufacturing conversion difficulties, or an increase in traffic casualties as a result of implementing this highly effective energy conservation strategy.

Type and Size of Engine

As is clear from the extensive discussion of motor vehicle power plants in the first sections of this chapter, there are engine designs available that can potentially result in increases in energy efficiency of up to 50 percent over standard pre-emission-controlled automobile engines, meet stringent pollution standards, and provide acceptable performance. These engines can and should be refined and employed. Their use will result in vast energy savings and pollution control gains.

The only further point to be made here is that relating to the effects of engine size. No engine—for that matter, no machine—is equally efficient at all rates of utilization up to its maximum technical capacity. Rather, engines are typically designed for maximum efficiency at a particular point or within a particular range of load. Therefore, in order to achieve the maximum efficiency of which the engine is capable, an engine should be chosen whose design range, so to speak, is consistent with the usual rate of output required for the given vehicle.

For this reason, the frequent use of vastly oversized (i.e., overcapacity) engines in domestic U.S. motor vehicles is an energy, money, and resource wasting practice. It makes absolutely no sense to fit an automobile, for example, with an engine capable of propelling the vehicle at speeds of more than 120 mph if the car cannot be driven legally at speeds in excess of 65 to 70 mph (even for short periods) and would not, in any case, be driven faster than 80 to 85 mph by more than a fraction of a percent of its potential owners.

The argument is often made that this oversizing of engines is required to produce that burst of speed required for safe passing and for emergencies. However, some types of engines can be designed to tolerate overcapacity bursts of power output for short emergency periods without providing for the capability to maintain that power for extended periods of time. And we have seen that hybrid engines may be designed in such a manner as to allow the basic engine to operate in its design range consistently with short power bursts provided by a small supplemental engine. This boost could also be provided by a supercharger. By one estimate, such a system could increase fuel economy by 15 percent, even without any change in vehicle weight.[29]

Thus, there is no legitimate technical reason for the installment of excessively large motor vehicle engines. A more rational matching of engine size with vehicle operational requirements would certainly conserve considerable energy without sacrificing quality.

Speed of Operation

The amount of (kinetic) energy that must be transferred to accelerate a vehicle of a given mass (hence weight) to its operating speed increases as the square of that speed. Furthermore, as operating speed increases above 30 to 40 mph, aerodynamic drag (i.e., the resistance of the air to the vehicle motion) becomes increasingly significant for vehicles of standard design. Thus, theoretically the energy requirements for motor vehicle operation are even more sensitive to variations in speed than to variations in weight.

Empirically, the decrease in the fuel economy of a typical automobile at the upper ranges of its speed is so sharp that an increase from 50 mph to 75 mph would result in almost doubling fuel consumption.[30] Even for cars with excessively large engines, the reduction of highway speed limits from 65 to 55 mph could be expected to increase motor vehicle fuel economy by at least 10 percent.[31] But, before such gains in energy efficiency can be registered, the lowered speed limits must be observed—which they clearly are not. Changing the signs does not affect fuel economy, changing the actual speeds at which vehicles are operated does.

It is sometimes claimed that decreasing the operating speeds of trucks actually reduces their fuel efficiency. Since transmissions are designed to optimize efficiency at the normal range of speeds, there may in fact be some reduction in fuel economy if speeds are lowered without adjusting transmissions. Certainly the full potential savings in fuel consumption will not be achieved unless transmissions are modified appropriately. But if they are, then reduced speed will increase energy efficiency.

Just as the lowering of operating speeds reduces the energy that must be transferred to the vehicle for normal operation, the energy that must be absorbed by a vehicle in collision is also reduced if the collision occurs at lower speed. Thus, all things being equal, the damage sustained by the vehicle and its occupants should be less severe in lower-speed collisions. Furthermore, because reduced speed increases the time available for reaction to emergency situations and reduces the distance required for braking, it is easier to avoid collisions entirely. Therefore, reducing vehicle operating speeds also conveys substantial safety advantages.

The safety effects of reduced speed were spectacularly demonstrated by the substantial reduction in traffic fatalities and the severity of traffic-related casualties during the period in late 1973 and early 1974 when the lowered highway speed limits were being widely observed under the pressure of the crisis gasoline shortage. Of course, this reduction in the carnage on the nation's highways was always achievable with a reduction in observed speed limits. It is an interesting comment on our national sense of values that we were willing to observe lowered speed limits to save gasoline but not to save lives.

The extension of travel time resulting from reducing average speed from 65 to 55 mph will be relatively small. The vast majority of automobile travel involves trip lengths of under 50 miles, yet even for a 50-mile trip, trip time would only be increased by about 8 minutes.[f]

Air and Road Resistance Parameters

The so-called aerodynamic profile of a vehicle is the shape that the vehicle presents relative to the wind—in particular, relative to its direction of motion. This profile is an important determinant of the amount of resistance offered by the air to the vehicle's motion. The more streamlined the shape of the vehicle, the more it is able to slice the air without generating undue disturbance, and hence, the less resistance it will encounter to its movement. Since the force represented by this air resistance is one of those that must be overbalanced by the engine to maintain velocity or to accelerate the vehicle, the more streamlined the aerodynamic profile, the less fuel will have to be expended by the engine.

In very crude terms, the more box-like the shape of the vehicle, the poorer its aerodynamic profile; the more cigar-shaped the vehicle, the better its profile. Of course, there must be a compromise between the vehicle shape which would minimize air resistance and that which would maximize the ratio of useful internal space to total volume, therefore aiding in the minimization of vehicle weight. The faster the vehicle is to move, the more the concerns of energy conservation would call for the shape to approximate the optimal aerodynamic profile, and thus, the further the design will be from that which maximizes the useful space ratio. This, then, is another reason why holding down vehicle operating speeds saves energy.

Current automobile and truck designs do not take these shape considerations sufficiently into account. In automobiles, particularly, these considerations are so subordinated to those of style that the resultant shapes often have neither good aerodynamic profiles nor high ratios of useful space to total volume.

Aside from air resistance, the vehicle must overcome so-called rolling

[f]Some additional energy savings may be achieved by reduced speed limits to the extent that they discourage some of the more marginal uses of automobile travel.

resistance. Reducing the amount of friction between the tire and the road will reduce this motion-retarding force, but of course this friction cannot be reduced too much because the ability of the tire to grip the road is crucial to efficient propulsion as well as effective vehicle control. No one who has ever driven an automobile or truck on an icy road will have any trouble understanding this.

The key to optimizing this frictional force lies both in the design of the tire and in the design of the road surface, but the latter will not be considered here. Of the two major tire designs currently available, the radial tire comes much closer to being optimal from an energy standpoint than the bias ply tire commonly used in the U.S. The radial tire has made a significant penetration into the U.S. market in recent years, and its widespread adoption would certainly be energy conservative.

It is reasonable to expect that a combination of improved aerodynamic form, the switch to radial tires, and proper tire inflation would produce a 10 to 15 percent improvement in overall fuel economy.

Braking

Presently, nearly all motor vehicles are fitted with frictional braking systems of one or another design. Such systems accelerate the increase in entropy by converting high-grade kinetic energy into low-grade heat, which cannot be readily recovered or re-used. By one estimate, the amount of energy that the brakes of an automobile dissipate is about one-tenth of the total work done at the drive wheels in suburban driving. The energy-loss for urban driving is more than three times as great.[32] We are thus talking about a considerable wastage of energy.

A *regenerative braking system* is any system that removes kinetic energy from the wheels of the vehicle and stores it in a form that can then be reconverted into kinetic energy at the drive wheels when reacceleration of the vehicle is desired. One possible regenerative braking scheme would use small electric generator–motors at each wheel. These would be operated as generators when braking was required, converting the energy of the spinning wheels into electricity which would then be used to charge a battery. When reacceleration was required, the battery would discharge into the electric devices now operated as motors which would then drive the wheels. Another regenerative system would use one or more flywheels as the energy storage devices, transferring the kinetic energy from the vehicle wheels to the flywheel during braking and reversing the process during reacceleration.

Batteries of the type now available for the first scheme described would return only 50 to 75 percent of the charging energy put into them and are relatively heavy.[33] Since their extra weight must be carried at all times, thus reducing vehicle energy efficiency, while their deceleration energy recycling feature operates only part time with restricted efficiency, it is possible that this system would only result in significant net energy savings if used in vehicles driven primarily in urban areas.

With the development light-weight, high tensile strength (fiber composite?) flywheels, much greater energy storage-retrieval efficiency should be achieved relative to weight, and hence the full potential of the flywheel regenerative braking system realized. For that matter, the development of high energy density–low weight electrical batteries would have a similar effect for the first system discussed.

If regenerative braking systems were used in place of frictional braking, a large fraction of the considerable amount of energy now wasted in the process of deceleration would be recovered and recycled. Such systems would be at least as effective in slowing or stopping the vehicle and would result in a significant reduction in total vehicle fuel consumption by reducing the amount of new energy that would have to be injected. Relatively simple, current technology designs would undoubtedly achieve net energy savings if employed in urban and possibly suburban vehicles. However, directing a substantial research and development effort to the refinement of such systems and, particularly, to the crucial development of high-efficiency, light-weight energy storage devices would seem to be a wise energy conservation strategy.

Transmission Design

The purpose of the transmission is to transfer the power developed at the engine to the vehicle's drive wheels. By varying the torque/speed ratio, the transmission can vary the amount of power delivered to the drive wheels at any given engine speed. Present transmissions, both automatic and manual, accomplish this variation by using a system of discrete gears. In low gears the wheels turn fewer revolutions relative to the engine speed than when the high gears are engaged. Thus low gears produce a more efficient coupling of engine and drive wheels when power is needed, say, for climbing hills, and high gears produce a more efficient coupling when speed is needed, as in highway driving.

Because they have a relatively small number of gears, present transmissions cannot vary the torque/speed ratio continuously but rather only in discrete steps. As a result, when the vehicle is being accelerated, full engine power is transmitted only at a small number of peak points (corresponding to the number of gears) rather than being available throughout the entire acceleration process. Thus a greater amount of engine power, and hence energy, is lost than would be lost if the transmission were continuously variable so that the optimum torque/speed ratio could be achieved at every point of the acceleration process.

A continuously variable transmission would allow a more efficient engine-to-wheels energy transfer than present, discrete transmissions. Fuel economy is enhanced in two ways by such a transmission: first, the same acceleration can be achieved with a smaller, lighter engine; and second, the engine can be operated in its optimally efficient design range during normal driving because the transmission better matches engine power with power demand.

There are a number of continuously variable transmission designs in existence coupled with appropriate control systems, and they should be thoroughly investigated.[34] An effective, reliable continuously variable transmission could potentially increase motor vehicle energy efficiency by up to 20 to 25 percent.[35]

Accessories

Accessories that require power input, such as air conditioning, power steering, power brakes, fan, power windows, etc., draw that power from the engine and thus increase the vehicle's fuel consumption per mile. There are two ways to approach the accessory energy problem: (1) eliminate the accessory; and (2) couple it more efficiently to the engine.

Certain accessories, most notably power steering and power brakes, are only necessary, in any sense of the word, to aid in the effective control of large, heavy vehicles. Therefore, when automobile weights are substantially reduced, as previously recommended, the need for these accessories will be terminated. Resultant fuel savings will likely be in the neighborhood of a few percent.

Automatic transmissions, as currently designed, impose a fuel penalty of 3 to 5 percent in intercity driving and 10 to 15 percent in urban driving, as compared to manual transmissions.[36] Although continuously variable transmissions are fully automatic, they are, as previously discussed, far more efficient than even currently available manual transmissions. However, until and unless such transmissions are made readily available, at least some gain in fuel economy can be achieved by using manual transmissions of present design. They are not difficult to use and they impose no compensating penalties.

Air conditioning increases fuel consumption by 10 to 15% in urban driving and by perhaps as much as 7 percent intercity.[37] The apparent frivolity of this accessory would seem to imply that the best approach would be to eliminate it. However, motor vehicle air conditioning not only greatly increases the comfort of the vehicle's occupants but may even be viewed as a safety feature. The greater thermal comfort, elimination of wind blast inside the passenger cab, and noise abatement made possible by air conditioning are all very important to the reduction of driver fatigue, irritability, and aggressiveness—all of which are of particular importance during extended periods of driving. Thus it is reasonable to consider air conditioning a worthwhile, though far from vital, accessory for both cars and trucks.

Fortunately, there are fairly straightforward ways of reducing the fuel penalty associated with air conditioning. Units may be designed that operate off the waste heat generated by the engine, or a small continuously variable transmission could be used to couple the engine to the air conditioning compressor and fan. Accessories driven directly by the engine (as they typically are) must have large enough capacity to perform effectively at low engine speeds and therefore waste energy at higher speeds. Interposing a continuously variable transmission (or for that matter any reasonable approximation thereof) should have the same type of efficiency-increasing effect for essentially the same reasons as the use of such a transmission has when interposed between the engine and the drive wheels. This coupling approach should be usable not just for air conditioning but for all energy-using accessories and could result in fuel savings on the order of 3 to 7 percent.[38]

Accessories like power window, power seats, etc. have no place in an automobile or truck driven by a person with anywhere near normal human physical capabilities. They are truly frivolous in that they have trivial affect on comfort, performance, safety, etc. and therefore must be viewed as purely wasteful of energy.[g]

Summary and Conclusions

The automobile and its freight-carrying counterpart, the truck, clearly dominate all other freight and passenger transportation modes in the United States. Unfortunately they are currently (with the exception of the airplane) the most energy intensive modes as well. We have considered a variety of ways in which the design and operation of these vehicles may be altered in order to greatly increase their energy efficiency.

Seven major classes of engines suitable for use in powering motor vehicles were evaluated. These included the Otto cycle engine (the presently standard gasoline-powered, piston-equipped automobile engine), the diesel (the presently standard large truck and bus engine), the Wankel rotary, the gas turbine, the Stirling, the Rankine, and electric engines. Hybrids were also considered.

The Otto cycle internal combustion engine was found to be an inherently dirty, energy inefficient engine. In general, the Stirling class of external combustion engines seem the most promising for general motor vehicle use because of its combination of high energy efficiency, low pollution, and good performance. Electric engines may be valuable for powering smaller, urban vehicles, provided advances are made in battery development. Rankine engines may prove most desireable for large, stop-and-go vehicles, such as urban buses.

---

[g]One possible exception would be automatic door locks or power rear windows in a taxicab.

In the short term, considerable improvement in the fuel efficiency and reduction in the emissions of Otto engines can be achieved by making use of the stratified charge concept currently in very limited use. Hybrid engines, built with off-the-shelf technology and materials could be incorporated into motor vehicles with little lead time. By using a smaller internal combustion engine operated at its most efficient range with a small supplemental engine providing acceleration boost power, fuel economy could be considerably enhanced and emissions reduced.

Vehicle weight was shown to be of great importance to energy efficiency. U.S. automobiles have tended to be increasingly oversized and overweight. A reduction in weight of the average U.S. car to 2,500 pounds would, by itself, increase automotive fuel efficiency by 30 to 40 percent, resulting in a reduction in total U.S. energy consumed for all purposes by as much as 5.5 percent and a reduction in total United States petroleum consumption of up to 11.5 percent. All this could be achieved within the context of vehicle designs currently in production without any sacrifice in the standard of living and even without elimination of luxury vehicles. Furthermore, small cars, in and of themselves, do not constitute a safety hazard—in fact, their greater maneuverability may increase safety.

Considerable increases in fuel economy could also be achieved by reducing the size of automotive engines to provide a closer match between real vehicle performance requirements and engine capabilities. Reduction in actual vehicle operating speed, as opposed to posted speed limits, has safety as well as energy advantages. Improving the aerodynamic profile of vehicles will reduce the air resistance, and improved tire design (e.g., radials), the rolling resistance that the vehicle must overcome thus reducing energy requirements per mile.

There is substantial fuel economy advantage to be gained by alterations in the design of both braking and transmission systems. Replacing frictional brakes by a regenerative braking scheme would recover and recycle energy that is now dissipated as heat, thus reducing fuel burning requirements. The use of continuously variable automatic transmissions in place of standard discrete-gear automatic or manual transmissions could increase the energy efficiency of automobiles and trucks by up to 25 percent.

Some fuel-consuming automotive accessories, mainly power steering and power brakes, will no longer be needed in vehicles of reduced size and weight. Others, like air conditioning, which do have real comfort and even safety implications, can be made more energy efficient by superior coupling of these devices to the engine. Finally, the more frivolous accessories, like power windows and power seats, are not worth retaining. Their elimination would hardly deal a staggering blow to the quality of these vehicles.

The improvement in the energy efficiency of automobiles and trucks that could be achieved by adopting the proposals set forth here is truly enormous. An immediate increase of 20 percent in the fuel economy of the typical new automobile is possible. In the short run, this saving could be increased to 33 to 50 percent or more. Within 5 to 10 years, automobiles and trucks that are at least twice as energy efficient as present day vehicles (and perhaps much, much more) could be widely available. And such vehicles can be far less polluting and even far more safe than current models. There is no reason to count the automobile or the truck out just yet. But there is also no reason of economics or engineering to accept the sacrifices in air quality, resource depletion, and even national sovereignty implied by continued widespread use of current motor vehicle designs.

Appendix 6A

Description of Alternative Engines

The Otto ICE. The operation of the standard piston-equipped, gasoline-powered internal combustion engine (ICE) approximates the so-called idealized Otto cycle. For this reason, it will be referred to as the *Otto ICE.* The Otto ICE involves a four-step power generation process:

1. An explosive mixture of air and gasoline vapor is drawn into the cylinder through an intake valve located at the top of the cylinder.
2. The valve is closed and a piston rises in the cylinder compressing the air gasoline mixture.
3. A spark ignites the compressed mixture and a rapid, explosive combustion takes place driving the piston down.
4. An exhaust valve opens and the pressure within the cylinder rapidly drops to atmospheric as the rising piston forces the exhaust gases out through another opening at the top of the cylinder.

The reciprocal piston movement produced by the exploding gas-air mixture is translated into rotary motion which is then transmitted to the drive wheels of the vehicle.

The fact that the Otto ICE extracts fuel energy by a series of timed explosions rather than by controlled combustion makes it an inherently dirty and inefficient engine. In order to approximate complete combustion of the fuel alone, a thoroughly mixed fuel–oxydizer charge must be burned under conditions of carefully regulated temperature and pressure. This degree of control is not likely to be attained under the extremely rapid explosive combustion conditions of the Otto ICE.

Incomplete combustion has two important consequences. First, part of the fuel energy will not have been extracted by the combustion process. To the extent that this energy is discarded (as is typically the case), energy resources are wasted. Second, the exhaust products of incomplete combustion are often dangerous pollutants. The three major air pollutants that are primarily due to fuel burning for transportation are carbon monoxide, hydrocarbons, and nitrogen oxides. The first two of these are directly due to incomplete fuel combustion and would be converted to carbon dioxide and water by more complete burning. The third is the result of the burning of nitrogen present with oxygen in the air mixture used as an oxydant.

The *stratified charge engine,* a relatively minor modification of the standard Otto ICE, has produced very considerable improvements in the effectiveness of the combustion process. Here two distinct combustion sectors are used in the cylinder, one containing a richer gasoline-air mixture than the other. The sector richer in gasoline is centered on the spark plug so that the combustion process will start easily. As the burning proceeds, the second sector of the cylinder is encountered and the leanness of the fuel-air charge there allows a more complete combustion to take place. The original system used an extra valve to produce this nonhomogeneous charge, but later systems achieved further simplifications including the redesign of the tops of pistons so as to create the desired nonhomogeneous charge distribution by producing appropriate aerodynamic conditions.[1]

The Wankel ICE.[2] Of the perhaps 30 to 40 types of rotary engines in existence, the one that has received the most attention is the Wankel internal combustion engine. Only one version of a class of rotaries known as *eccentric-rotor engines,* the Wankel is an extremely simple power plant. It has only two primary moving parts—the rotor and the drive shaft.

The rotor, which is shaped like a triangle whose sides bow outward (i.e., are concave to the rotor's center), is pierced at right angles by a cylindrical driveshaft. The center of the circular cross section of the shaft is not the same as the center of the rotor, hence the rotor is offset, or eccentric, to the driveshaft. The rotor moves in one direction inside a chamber shaped so that all three points of the triangular rotor sweep along the chamber walls, always in contact with the walls except where openings in the chamber wall are deliberately provided. Because of the peculiar trochoidal shape of the chamber and the fact that the rotor is offset on a driveshaft which itself is centered in the chamber, the space between any one side of the triangular rotor and the chamber wall alternately increases and decreases in size as the rotor spins.[a]

At the point where this space is the largest, a stationary intake port is provided and the fuel-air mixture flows into this area. With the turning of the rotor, the area between this side of the rotor and the chamber wall becomes smaller, compressing the mixture. The continued rotation soon brings this compressed fuel-air charge to the point in the chamber wall where a spark plug is located. A spark from the plug ignites the mixture and the resulting explosion drives the rotor to the point where a stationary exhaust port is provided, allowing the gases to exit from the chamber. Since the spaces between each side of the triangular rotor and chamber wall go through all of these phases one after the other, each phase continuously takes place somewhere within the cham-

[a]The chamber shape is roughly that of the outer boundary of the figure made by two intersecting circles. It approximates an oval which is pinched in at the midpoint of each of its long sides.

ber.[b] Thus, by clever specification of the shape and relative position of the rotor and chamber, the same intake-compression-explosion-exhaust cycle produced in the Otto ICE by reciprocating pistons and valves is produced in the Wankel, essentially by geometry.

The Diesel ICE. Like the Otto ICE, the diesel is an internal combustion engine that contains reciprocating pistons enclosed in cylindrical chambers. In fact, its operation is very similar to that of the Otto cycle engine.

In the diesel cycle a blower forces high-pressure air into the cylinder through an intake port located in the side of the cylinder wall. This air not only provides part of the fuel-air charge but also helps force exhaust gases from the previous cycle out through exhaust valves at the top of the cylinder. The exhaust valves close as the piston rises in the cylinder compressing the air. This compression so raises the temperature of the air that when fuel oil is injected at the top of the compression stroke, it ignites and burns without requiring a sparking device. The combustion is not as rapid as in the gasoline-fueled Otto ICE, and the first part of the power stroke, in which the piston is forced down by the expansion of the burning gases, proceeds at nearly constant pressure. When the piston travels downward it passes the open air-intake port and fresh high-pressure air is blown in as the exhaust gases exit through the open exhaust valves.

The fact that there is no fuel in the cylinder of a diesel engine during compression eliminates the possibility of pre-ignition, a condition in which the fuel-air charge ignites prematurely and thus not at the part of the cycle at which optimum efficiency in energy transfer is achieved. Furthermore, the absence of fuel during compression also allows the achievement of a high *compression ratio* (the ratio of maximum to minimum volumes of air in the cylinder during the cycle), and higher compression ratios imply higher engine thermal efficiencies.[3]

The Gas Turbine Engine.[4] Unlike all of the previously discussed internal combustion engines, the gas turbine fires continuously rather than intermittently. Air is first drawn in through an inlet diffuser. Then it passes through a compressor that produces a sharp increase in pressure along with a more moderate increase in temperature (in approximate accordance with the equation of an ideal gas). The air, now at high pressure, enters a combustion chamber into which fuel is sprayed by a nozzle and the resulting fuel-air mixture is burned producing an intense flame. Though the temperature of the gas, of course, increases sharply in the combustion chamber, its velocity remains essentially unchanged, and its pressure even drops slightly. Now the gas passes through a turbine (a wheel or set of wheels with curved vanes–similar in concept to a

[b]The sparking is, of course, intermittent.

fan) causing it to rotate rapidly. The turbine is attached to a driveshaft which transmits power from the turbine to operate the air compressor and to drive the vehicle's wheels. The gases then exit through an exhaust nozzle.

In aircraft applications of the gas turbine engine, the net force (thrust) which moves the aircraft results from the difference in momentum between the inlet and exhaust gases. Accordingly the exhaust nozzle is designed to vastly increase the velocity of the gas, and the turbine itself is essentially used only to drive the compressor. In automotive application, as much of the kinetic energy of the gas as possible must be absorbed by the turbine for transmission to the drive wheels (as well as whatever is needed to drive the compressor), because developing motive force through high-velocity exhaust blast is simply not practical for such land vehicles.

Initially, one of the serious problems of the gas turbine as a motor vehicle power plant was precisely the problem of superheated, high-velocity exhaust. The heat problem was largely overcome by installing rotating heat exhangers (regenerators) between the turbine and the exhaust port. These transferred heat from the exhaust gases to the air coming from the compressor, thus improving combustion and therefore fuel economy as well as reducing exhaust heat.

The Stirling ECE.[5] The most striking difference between the Stirling engine and those previously discussed is that it is an external combustion engine. This means that the combustion process occurs outside the engine itself in a burner that may be separately designed for maximally efficient continuous combustion of whatever fuel is to be used. Since the engine operates solely on the basis of heat drawn from a heat source, the type of fuel burned is wholly irrelevant to its efficient functioning. For that matter, the heat need not even be generated by combustion. The use of solar heat, for example, is also theoretically compatible with the Stirling ECE.

The basic working elements of a Stirling engine are two spaces whose volumes may be changed (typically by pistons) and are separated by a regenerator (a device that may be described as a "thermal sponge," alternately absorbing and releasing heat). Heat is continuously supplied to one space from an external source, while the other space is continually cooled by an external cooling device, thus creating a temperature differential. A working fluid (air or some other gas) is present.

In the conceptually simplest version of a Stirling engine, the two spaces are at opposite ends of a cylinder whose central section contains a matrix of finely divided metal strips, which serves as the regenerator. A piston is located at each end of the cylinder and is free to move from its end of the cylinder through the space up to the edge of the regenerator matrix. Heat is supplied to the top space in the cylinder, called the *expansion space,* through tubes that transfer heat from the external burner. The bottom space, called the *compres-*

sion space, is cooled by cooling tubes that operate like the radiator of a standard automobile. Gas fills the cylinder between the two pistons which face each other.

The cycle begins with the top piston at the edge of the regenerator (i.e., as far down as it can go) and the bottom piston at the lower end of the cylinder (i.e., as far down as it can go). The gas is at the lowest temperature it will have during the cycle (since it is in the cooled compression space) and also occupies the largest volume it will occupy during the cycle. Since the pressure exerted by a gas varies directly with temperature and inversely with volume (according to $PV = n\mathbf{R}T$, the equation of an ideal gas), the gas has its lowest pressure of the cycle here.

The bottom piston moves upward, compressing the gas against the unmoved top piston and thus raising both its pressure and its temperature. The gas is forced through the regenerator where it picks up heat stored there in a previous cycle, causing its temperature and thus its pressure to increase further. The heated high-pressure gas forces the top piston upward to the top of the cylinder. The bottom piston is now at the edge of the regenerator (i.e., as high as it can go) and the top piston at the upper end of the cylinder (i.e., as high as it can go). Thus the volume of the gas has expanded back to its original uncompressed volume. Since volume has increased during the expansion process, the pressure of the gas will have fallen. However, since the gas is now in the heated expansion space, its temperature is higher than when the cycle began, so that even though it occupies the same volume, its pressure is higher than at the start of the cycle.

Both pistons move simultaneously down to their original positions, causing the gas to pass through the regenerator, which absorbs heat from the gas and stores it for the next cycle. The volume of the gas does not change in this final phase because of the simultaneous movement of both pistons in the same direction. The work done by this cycle is done during the expansion process. The reciprocating piston movement is mechanically transformed into the rotation of a driveshaft which transmits power to the drive wheels.

In multicylinder versions of the Strirling ECE, the cylinders may be linked sequentially, with the working fluid free to move between the space at the bottom of one cylinder and the space at the top of the next cylinder. In such a Rinia arrangement, the regenerators are located in special casings between the cylinders. Thus the movement of a piston in response to the high pressure of the gas during the expansion phase itself compresses the gas relative to the next cylinder and piston, allowing a halving of the number of reciprocating elements.

The Rankine ECE. Like the Stirling engine, the combustion associated with the Rankine engine does not take place in the engine itself. In modern versions of the engine, a working fluid (e.g., water) is sealed within a closed system. The fluid, initially a liquid, is pumped into a boiler where it is heated by combustion gases to a hot high-pressure vapor. The vapor then flows through an intake valve into a cylinder in which it expands against a piston, thus increasing in volume and accordingly dropping in pressure and temperature. The low-pressure vapor passes through a regenerator where part of the heat is recovered, then through a radiator-like heat exchanger where it is condensed to a liquid as more of its heat is removed. Liquid from this condenser is pumped back to the boiler by a pump powered by the engine itself, and the cycle is repeated.

The work, of course, is done in the process of expansion of the vaporized working fluid against the piston. Again, the reciprocating piston motion is translated into rotation by a crankshaft and the rotary motion mechanically transmitted to the vehicles drive wheels. Some considerable success has reportedly been achieved in reducing internal engine corrosion and mitigating the problem of freezing of the working fluid by shifting from the use of water (as in early designs) to the use of various inexpensive organic working fluids.[6]

The Electric ECE. Efficient electric motors have been in existence for a very long time. Large and small electric motors are commonly used for a variety of purposes. It is not the design of the electric motor itself that poses the problem, rather it is the design of either an efficient mode of connection between the electric motor and the central power station compatible with a mobile vehicle, an efficient portable storage device or an efficient compact light-weight generator, that is the problem.

There are a number of electrically powered motor vehicles now in common operation, but they are for the most part low-speed, low-performance vehicles (e.g., golf carts, fork lift trucks, small delivery vans), relying on packs of lead-acid batteries as storage devices. These batteries are essentially the same as those typically used in automobiles to power starter motors and accessories. Their design is virtually the same as that of batteries used in the electric cars produced at the start of the twentieth century![7]

The storage device route seems the most promising at the moment, but it will not be possible for electric motors to be considered a serious alternative motor vehicle power plant for general use unless a relatively inexpensive, high-energy, high power-density, light-weight storage device is developed. At the moment there appear to be two possible approaches to this problem: concentrate on developing commercially feasible versions of some of the advanced energy storage devices developed in connection with the space program (currently extraordinarily expensive); or produce a modernized, high-efficiency version of a mechanical energy storage device whose basic technology has been well-known for centuries–the flywheel.[8]

The flywheel is one of the oldest inventions of humankind.[c] It is simply a wheel mounted on an axle and allowed to spin freely. As energy is added to

[c]The potter's wheel, mentioned in the Old Testament, that was kept spinning by intermittent kicks of the potter's foot operated by flywheel action.

the wheel, the wheel spins faster, storing the energy imparted to it as mechanical energy, until frictional forces eventually cause the energy to dissipate or until the energy is purposely removed by attaching a load to the wheel's shaft. Since the flywheel stores *kinetic energy,* the amount of energy stored at any given time depends positively on both the mass of the wheel and the *square* of its speed of rotation. It would seem any amount of energy could be stored in the wheel by simply spinning it faster, but in practice the stresses built up by spinning the wheel faster and faster would eventually exceed the tensile strength of the material and the wheel would fly apart.

For equal tensile strengths, lighter materials make more efficient flywheels because they will tolerate higher speeds (since they have lower density) before the stresses build up to the limit and thus can store equal amounts of energy within lower mass, achieving a higher energy storage to weight ratio. Modern fiber composite materials combine light weight with high tensile strength and therefore would potentially make excellent automotive energy storage devices to be used in conjunction with electric vehicles, when encased in small evacuated chambers to reduce air friction.

The design and operation of such an electric vehicle would be as follows: the electric motor running on power from an electric wall outlet would be used to spin up the flywheel, i.e., to "charge the battery," as it were. Then the energy stored in the flywheel (carried by the car) would be used to generate electricity and supply it to smaller electric motors, say, one on each wheel, thus driving the car. The flywheel could be recharged in a few minutes and, if properly designed, would require months to run down from full charge when the automobile is not being used. Furthermore, this design is easily compatible with a regenerative braking system that would use the wheel motors as generators when the car is decelerated or driven downhill, putting energy back into the flywheel so that it would be available for subsequent use. The range of the car could be increased by 25 to 50 percent with such a system, and considerable energy savings per mile achieved.

Thus the electric motor is a viable alternative to the other automotive power plants discussed, if such inexpensive, high-energy storage devices as the fiber composite flywheel can be perfected.

7 Land Transportation: Railroads, Bicycles, Buses, and Urban Systems

In the previous chapter we focused attention on the improvement of the energy efficiency of the two most popular but least efficient land transportation modes, the automobile and the truck. Here we consider the changes necessary to make the other, more efficient means of land transport more attractive. Although further improvement of the energy efficiency of these modes is also highly desirable, the energy savings achievable by accomplishing a shift in the traffic distribution in their favor are likely to be of considerably greater magnitude. The primary means for achieving this shift are the improvement of the speed, comfort, convenience, and coverage of these alternative modes without an offsetting increase in their relative expense. This may be done by a combination of design changes, improved operating procedures, and increased coordination.

Railroads

The energy efficiency of the railroad has increased continually over the past three and one-half decades; at the same time, its share of passenger and freight traffic was undergoing a precipitous decline. The largest jump in railroad energy efficiency came during the 1950s with the nearly complete replacement of open-cycle steam locomotives by diesel-electric engines. As a result, whereas steam was the principle motive power for the railroads in the first half of the twentieth century, more than 99 percent of the locomotives in service in the United States in 1974 were diesel-electric.[1] In other parts of the world where diesel fuels have not been as cheap as in the United States, there has been a growing use of pure electric locomotives. It is therefore of some interest to consider the relative merits of these two modern railroad power systems.

The Diesel-Electric versus the Electrified Locomotive[2]

In the diesel-electric locomotive fuel is burned in an on-board diesel engine that is used to drive an electric generator. The current from the generator is then fed into electric traction motors mounted on each axle. The power system for the pure electric locomotive is somewhat more elaborate. Electricity is generated at a central power station and transmitted via wires suspended from poles near the tracks (called *catenary transmission lines*) or via an electrified third rail. The locomotive carries a pantograph or other current collecting device that draws the electricity from the transmission lines, making continuous contact with them as the train moves. The electricity produced by an external power generating system is thus gathered in by the locomotive and directed to electric traction motors mounted on each axle.

The diesel-electric system is far less capital intensive, since the self-contained power unit eliminates the need for miles and miles of transmission lines and their supporting systems. It also consequently avoids the energy losses associated with such lengthy transmission lines. On the other hand, the diesel-electric must haul a large on-board engine, along with the fuel it requires, plus a sizeable cooling system, all of which burdens the locomotive with far more weight than that carried by the electrified locomotive. As we have seen, excess weight always results in an energy penalty.

Beyond this, the size of the cooling system required to remove the buildup of waste heat from the engine becomes so large as the horsepower of the diesel is increased that it becomes a critical design limitation.[3] As a result, though there are pure electric locomotives available in the U.S. capable of providing equivalent diesel engine output of 6,000 horsepower continuously and up to 10,000 horsepower for limited periods of acceleration, there are no single-engine diesel-electric locomotives available to the United States railroads that are capable of providing more than 4,000 horsepower.[4] This locomotive horsepower handicap is of great importance because it is critical to the hauling speed that the diesel-electric is capable of achieving and the trailing load (the weight of the train less the weight of the locomotive) it is capable of handling. The former consideration is important to the attractiveness of the railroad as a transportation mode, while the latter greatly influences energy efficiency.

Furthermore, in present designs, diesel-electrics have experienced serious reliability problems. However, in a survey conducted by British Railways (reported in the December, 1972 issue of *Railway Gazette International*) only 25 percent of the failures experienced by diesel-electrics were due to the engine itself while about 65 percent were due to the cooling system and were often caused by nothing more sophisticated than simple leaks.[5] Nevertheless a failure is a failure, and the propensity of locomotive breakdowns to generate chain reaction delays by clogging a section of track makes the reliability problem especially severe. Hauling a back-up locomotive in case the primary locomotive breaks down would mitigate the reliability problem, but pulling around more than 200 tons of excess locomotive could impose a fuel penalty of up to 10 to 20 percent. Pure electric locomotives (without large cooling system requirements) may be a superior means of increasing reliability.

On the average, the energy efficiency of a typical 3,000 horsepower diesel-electric locomotive is probably 28 to 30 percent, including all losses between the fuel and the power delivered at the wheels.[6] The overall energy efficiency of an electrified locomotive is a bit harder to determine, because it depends so critically on the efficiency of the central power plant generating the electricity. If the central plant was fossil fueled, the overall energy efficiency of the locomotive would be 26 to 32 percent, depending on the age of the plant. If the central plant was hydroelectric, a locomotive energy efficiency of nearly 60 percent is likely.[a,7]

Regenerative braking is also generally easier to achieve with electrified locomotives because they do not need a capability for on-board energy storage in order to take advantage of this system. When descending a grade or decelerating, the electric traction motors are simply arranged to function as generators, converting the train's mechanical energy into electricity that can then be fed back into the external electric transmission lines (e.g., the catenary) for use at other locations of the railway or return to the power system. Thus the electrified locomotive can use the power system for storage rather than carrying a weight-increasing energy-storage device. Of course, regenerative braking could also be accomplished with an on-board energy storage device, as in a diesel-electric. It is just that the pure electric has this additional attractive option.

On the whole, on grounds of energy efficiency, performance, and reliability it would appear that electrified locomotives are superior to diesel-electrics for general railroad use, at least insofar as present designs are concerned. The energy advantage of pure electrics is marginal where power is generated at new fossil fuel plants, but overwhelming where hydroelectric power is available.[b]

Although this discussion has been limited to a comparison of the diesel-electric and the electrified locomotive, this should not be taken to imply that no other reasonable alternatives to these engines exist. For example, an improved closed-cycle Rankine engine may prove to be an attractive design. The closed nature of the cycle would eliminate some of the disadvantages of the open-cycle steam engines used prior to the success of the diesel-electric, including particularly the need to keep replenishing the water supply. The gas turbine engine is another real possibility, particularly for high-speed trains. Experimental versions of the gas turbines were put into use in France in 1966 and were being used for regular passenger service there by 1970.[8] The focus on the diesel-electric versus the pure electric then is due to their premier position among currently available power systems for general railroad use.

Inducing the Shift to Railroads

There is no significant cargo incompatibility between railroads and trucks insofar as freight traffic is concerned. That is, there is nothing in the nature of the cargo

[a]Efficiency with a nuclear generating plant is likely to be lower than even that of a fossil fuel plant. In addition, severe reliability problems commonly experienced by nuclear plants would likely more than undo the reliability advantages of the pure electrics.

[b]This suggests that the electrification of the railroads servicing the Northwest U.S. may be a wise energy policy.

itself that indicates which of the two modes is appropriate. Routing or scheduling considerations may sometimes point to a preference for truck shipment, particularly for short hauls, but on the whole most cargo is technically shiftable.

The greatest potential relative advantage of railroads in passenger transportation is for trips of intermediate length, say 100 to 500 miles. Because of their center-city-to-center-city capability, the total effective trip times achievable by efficiently run high-speed railways for such runs should be easily competitive with air travel. For short hauls, railroads are also potentially attractive along heavily traveled commuter corridors. It is even conceivable that they could offer real competition to the airlines on transcontinental hauls, but that is far less likely.

In order to fully exploit the potential attractiveness of energy efficient rail travel, three critical subobjectives must be achieved: reliability of service must be greatly improved, higher levels of speed and comfort must be attained, and the coverage of the rail net must be extended.

Improving Reliability. Throughout the early decades of widespread rail travel in the U.S., the railroads took pride in the tenacity with which they clung to their time tables. The clock was almost a symbol of railroading. Today you can still set your watch by the train timetables in much of Europe and in Japan, at least for the major lines. But in the United States being on time has become so unusual that many delays are scarcely even announced. What has happened?

After years of relative neglect and decline, the railroads now operate an aging and, in many respects, technically obsolescent rolling stock on a contracted, aging network of tracks. The financial deterioration of the railroads has often led to short-term cost-cutting reductions in maintenance that have further aggravated the physical deterioration of operating equipment.

The railroads have, along with much of civilian industry, suffered from the effects of the deterioration in civilian research and development capability that has accompanied the prolonged and continuing "brain drain" of much of the nation's engineering and scientific talent into military and space-oriented projects.[9] The consequent effects on the rail equipment supply industry have been such that even newly delivered rail equipment has suffered from severe reliability problems. For example, during 1969 the Penn Central Railroad found it necessary to have a 10 replacement cars available in order to keep an average of 18 cars operating on the New York–Washington Metroliner run; two of the cars newly delivered to the railroad that year has to be scrapped for spare parts, while 20 more cars were delivered with serious defects some 6 or more months after the agreed upon delivery date![10]

No matter how talented or well-intentioned the management, no matter how highly skilled or motivated workers, it is simply not possible to run an efficient, reliable railroad system under these conditions. There is no quick, neat, inexpensive solution to these problems of long-term, large-scale deterioration of the railway system. In the first place, a massive transfusion of capital is required to upgrade the quality and modernity of the rolling stock, the track, and other related equipments.[c] The application of this capital must be carefully monitored so that it is not largely absorbed in inflated prices for the same low-quality products or in the creation of bloated staffs of administrative employees. Secondly, increased research and development labor power must be channeled into the improvement of rail technology.

Neither of these steps will be easy. Both will be expensive. But both are highly energy efficient investments that will also tend to improve the standard of living by upgrading the quality of the nation's transportation infrastructure.

Increasing Speed and Comfort of Passenger Trains. In 1964 the Japanese National Railways opened the New Tokaido Shinkansen line and has been routinely operating passenger trains at speeds of 125 to 160 miles per hour ever since.[d] What is even more striking is that they did so at a profit, illustrating the value of high speeds to the commercial success of passenger rail service.[11] It should also be pointed out, however, that the Shinkansen is not only fast but reliable, quiet, relatively smooth riding, and equipped with comfortable seats.

Of course, as the speed of the train is increased, the energy required to propel the train increases even faster. Thus high-speed trains are inherently less energy efficient than standard trains. However, since standard railroads are some 200 to 275 percent more energy efficient than aircraft in passenger transport,[12] to the extent that higher-speed trains can effectively draw passengers predominantly from air travel, they can be considerably less energy efficient than conventional trains and still produce a net savings in energy consumption. It is important to collect and analyze data on the energy performance of these trains in order to evaluate at which speed the energy conservation gains from shifting air travel to highspeed trains tend to vanish (at least for present designs).

The diminution of energy efficiency with increasing train speed is greatly enhanced by air resistance—in fact air resistance may be the most important limiting factor in the achievement of high speed. Tests on the French TGV turbo train have shown that at speeds of 185 mph air resistance absorbs 95 percent of the traction power.[13] By implication, if the speed of this train were increased to about 250 mph, its energy consumption would increase by 250 percent,[14] quite possibly rendering it less energy efficient than an aircraft.[e]

[c]The word transfusion is used in place of infusion in order to highlight the fact that this flow of capital to the railroads must come from some other sector.

[d]France, Canada, Great Britain, West Germany, and Italy have also tested high-speed trains with design speeds ranging from 120 to 185 mph. By 1973 all but the West German and Italian trains were in operation, and those were scheduled for operation by 1975-1976. The U.S. Metroliners are designed for operation at speeds of 160 mph but cannot exceed 95 mph in operation because of track conditions. The power systems used in these high-speed trains include: electric, diesel/electric, turbo, turbo/hydraulic, and turbo/electric.

[e]The French tests also indicate that using steel wheels on steel rails results in essentially the same power requirement as magnetic-levitation vehicles and is far less power intensive than air-cushion vehicles.

As for the difficulties posed by aerodynamic shock waves generated by high-speed trains, there has been some success in mitigating this problem. The early results of the TGV tests show that the shock effects of the train when operated at 185 mph on either passing trains or on pedestrians near the track is no greather than that of an ordinary train operated at 125 mph.[15] The problem of shock when entering tunnels has apparently been successfully dealt with by the Japanese National Railways, since the Shinkansen line runs in and out of more than one hundred tunnels without any significant discomfort to passengers or operating problem for the train.[16] Italian State Railways is constructing specially shaped tunnel mouths to ameliorate the shock problem for its 160-mph train.[17]

Horizontal and vertical track stresses produced by properly designed 160-mph trains are reportedly no greater than those produced by the present 95-mph trains, but vibration has been a problem.[18] Nearly all of the world's high-speed trains designed for operation at speeds above 130 mph require substantial track modifications. Failure to upgrade track cripples the effort to provide high-speed service, as illustrated by the fact that the United States Metroliner must be operated no faster than 95 mph though its design speed is 160 mph. This again indicates the critical requirement for a massive capital transfusion into the United States rail system.

Thus, although there are still problems remaining, there is more than enough test evidence and operating experience to indicate the feasibility of running efficient high-speed trains. Attractive, comfortable interiors are easily compatible with the design operating characteristics of these trains. Furthermore, there is some real indication that such comfortable, fast rail service can be provided economically and consequently that the more energy efficient railroads can draw significant traffic from the less energy efficient airlines, improving the modal mix. It is, however, important to emphasize that if rail speeds are increased too greatly, the energy advantages of such a shift will evaporate. On the other hand, rail speeds do not have to match or even approach air speeds before railroads can provide faster city-center-to-city-center service.

Increasing Speed of Freight Trains. In 1974 the average freight train speed in the United States was about 20 mph, so that freight speeds are of a different magnitude than those of passenger trains.[19] Since aerodynamic drag does not tend to become significant until train speed rises beyond 40 to 50 mph, the primary problem in increasing freight train operating speed is the huge amount of energy transfer required to accelerate their enormous mass.[f]

As long as the rate of speed is held to relatively low levels, the energy effi-

---

[f]The average weight of a carload of freight was 58 tons in 1974, and the average freight train hauled more than 65 cars, implying that the total gross weight of the average U.S. freight train was over 4,000 tons in that year (see Bureau of Railway Economics, *Year of Railway Facts,* pp. 41–42).

ciency penalty for increasing train operating speed is probably not intolerably high. For example, on a flat track an increase in train speed from 30 to 40 mph may only result in about a 13 percent reduction in fuel efficiency.[20] But increases in operating speed may not really be the critical factor in increasing the speed of rail freight service.

According to the Association of American Railroads,

> . . .the average serviceable (freight) car spends about 12 percent of its time in road trains, loaded or empty. The remainder of freight car time is consumed mainly in loading and unloading operations, moving from place to place within terminals, being classified and assembled into trains, or standing idle during seasonal lulls in car demand.[21]

With the exception of the seasonal idle time and some of the intraterminal movement, all of these other activities are closely associated with the effective speed of rail freight. If the time required for loading, unloading, and assembly could be substantially reduced, it seems highly probable that the effective speed of rail freight service would be substantially increased.

It is also possible that the same scheduling, routing, and materials handling improvements that reduce their freight car dead time will simultaneously increase the reliability of rail service. Increased use of computers for scheduling and routing and increased containerization of cargo may be effective in both these respects. Properly designed freight containers can and do facilitate interconnection with truck service that may be required for initial gathering or final distribution of freight and/or with water-borne transport. In this way the efficiency and reliability of the entire freight transportation net may be improved.

Expanding the Rail Net. Between 1930 and 1972 the total mileage of railroad track operated in the United States declined by nearly 75,000 miles, a figure that represents more than 20 percent of the 1972 total.[22] This contraction has resulted in the effective destruction of much of this track, either through the physical deterioration, which proceeds with fair rapidity in disused track, or through physical removal or paving over.[g] There are no existing comprehensive studies of the condition of railroad trackage for the U.S. as a whole or of the cost of upgrading what is generally agreed to be a degenerating track system.[23]

Clearly, the railroads cannot under any circumstances compete for freight and passenger traffic over routes along which there is no usable track. And the more restricted the coverage of any mode of transportation, the less potential customers tend to think in terms of using that mode. That is, if there is effective rail service to say two out of five cities to which a given individual travels with

---

[g]The contraction of the rail net has had sociopolitical as well energy effects. For instance, one of the main grievances of the people of the Watts area in Los Angeles, which led to rioting in the late 1960s, concerned the unavailability of mass transit. And yet Watts was once a main terminus for an extensive California rail system.

variable frequency but all five are serviced effectively by the airlines, that individual may simply learn to think in terms of going by air rather than stopping to consider whether his or her next destination is serviced by rail as well. Therefore, the more restricted the rail net, the more traffice the railroads will tend to lose, even on routes they do still service.

Expanding the coverage of the rail net, or rather undoing its contraction, is an important part of improving the attractiveness of the railroads. It will be expensive, but if intelligently done it will likely prove to be a very effective investment in energy conservation.

Bicycles

The bicycle is by far the most energy efficient means of human transportation. It is not only much more efficient than any nonhuman-powered machine, it is even five times as efficient as walking.[24] A cyclist in good physical condition can achieve the equivalent of more than 1,000 passenger-miles per gallon.[25]

Apart from their nearly incredible energy efficiency, bicycles also generate no pollution as a result of their operation. Even their manufacture is far less resource depleting and pollution generating per unit than that of any other vehicle, because of their small size and simplicity. In addition, the human exercise that cycling provides conveys real health benefits under normal conditions, aiding in weight reduction, effective circulatory function, and improvement of muscle tone.

In 1971 about 18 percent of the total urban automobile passenger-miles involved trip distances of five miles or less, with an estimated 63 percent of this mileage being traveled during daylight hours in good weather conditions.[26] This would imply that a maximum of about 11 percent of urban automobile mileage could, on grounds of weather and distance, be replaced by bicycle travel. The actual substitutability is likely to be far lower for present bicycle designs because of unfavorable terrain, the necessity for carrying heavy packages (e.g., during food shopping trips), infirmities, etc. If it is arbitrarily assumed that 80 percent of this distance is not substitutable, then slightly more than 2 percent of total United States urban automobile passenger-miles would still be shiftable to bicycles.

It has been estimated that the energy required to grow, process, transport, and prepare the food needed to provide the incremental human energy burned up in the process of cycling amounts to some 790 Btu/mile.[27] This yields the equivalent, in terms of gasoline, of roughly 165 passenger-miles per gallon[h]—nearly 5.5 times as great as a generous estimate of 30 passenger-miles per gallon for the average urban automobile (with actual loading). Therefore, for each pas-

[h]This figure is substantially lower than the 1,000 passenger-miles per gallon primarily because it includes all of the energy inefficiencies involved in the growing, processing, and delivery

senger-mile shifted from automobile to bicycle the equivalent of roughly .025 gallons of gasoline are saved.[i]

If we follow the previously mentioned assumptions on the percent of urban auto travel substitutable by bicycles, in the sample year 1971, then the implied net fuel savings from this shift totals more than 1 billion gallons of gasoline per year (equal to an annual energy saving of about 130 trillion Btu's). This is enough gasoline to power a fleet of 1,000 full-sized United States cars the equivalent of four times around the world!

In fact though, the net savings would likely be considerably greater than this because the food calorie intake of people in the United States is for the most part already substantially in excess of their normal physiological needs. Therefore, it is improbable that this sort of a shift to bicycles would result in a net increase in United States food consumption so much as it would result in less of the present calorie intake being worn as excess body fat. If it were actually true that aggregate food consumption was not increased at all by this shift, the resulting energy savings would be on the order of 20 percent greater than the previous estimate. Therefore, if 20 percent of the less-than-five-mile, good-weather urban automobile trips were shifted to bicycles, net annual energy savings of somewhere between 130 to 155 trillion Btu's, or 1.0 to 1.2 billion gallons of gasoline equivalent could be expected.

Enhancing Bicycle Substitutability

The substitutability of bicycles for automobiles in urban travel, even for short trips, is limited by a number of important considerations. Speed, however, is *not* one of them. Although automobiles are obviously capable of reaching and sustaining much higher speeds than bicycles, under the conditions of traffic and limited space (for driving and for parking) common to urban areas, bicycles generally have an advantage insofar as effective trip speed is concerned. By way of illustration, in a study of the time required for completion of a series of 25 urban trips averaging 5 miles, bicycles were faster in 21 cases, cars in 3, and there was one tie.[28] In another experiment, bicycles were faster in 9 out of 10 residence to shopping center urban trips.[29]

Some of the considerations that do limit the substitutability of bicycles are excessively hilly terrain, bad weather, the need to carry cargo, and safety problems. It is probably not possible to overcome these problems completely by

---

of foods. In other words, the cyclist may be capable of going nearly 1,000 miles on the equivalent of one gallon's worth of Btu's in food calories, but something approaching 6 to 7 times the cyclist's direct energy intake was required to provide the food (under the present highly energy intensive system).

[i]Assuming, of course, that the shifting of this mileage results in the equivalent reduction in car trips and not merely in decreased car loading.

design improvements, but it is certainly possible to reduce the degree to which they constrain substitutability.

Overcoming Terrain Limitations. Anyone who has ever ridden a bicycle up even a relatively mild hill will appreciate the considerable incremental effort required to do so over what is needed on level ground. Therefore, if a bicycle were designed with some sort of booster device, to be used only when hill climbing, the terrain constraint would be somewhat relaxed.

The problem is to devise a booster that not only provides sufficient bursts of power but also is light-weight and small in size. In addition, it should neither cause any vibration or control problem for the rider nor result in undue noise, pollution, or energy consumption.

Since it is normal in hilly terrain to go down hill as well as uphill, it seems that the most appropriate booster would be a device that absorbs the energy in downhill runs, thus aiding in braking, and stores it for release when required for uphill runs. Such a device has already been discussed with respect to electric vehicles—namely, the flywheel. However, the need for light-weight, (say fiber-composite) materials is especially strong in this application. To make even more effective use of the flywheel, it might also be designed to absorb energy from road shocks. An experimental bike combining a flywheel with a movable (up and down) saddle has reportedly already been built.[30] In principle, the idea is intriguing because it could simultaneously increase riding comfort while adding energy to the booster. If properly designed, the flywheel could also be attached in such a way as to allow arm muscle power to be used to start it spinning before the cycle is even mounted, or to increase its energy while in motion. Once more, the importance of developing light-weight, high tensile strength materials that are relatively inexpensive (and not excessively energy intensive in manufacture) is highlighted.

The booster function could alternatively be performed by a small electric motor energized by a battery charged from wheel-mounted generators, but this arrangment may not be as energy efficient. At any rate, it too requires some technological development—in this case the development of a high energy density, light-weight electrical energy storage device.

Of course, a very small petroleum-powered engine could also serve as the booster. Aside from not being nearly as elegant a solution as either of the other two, there would be some pollution, nonhuman energy use, and most probably noise. Nevertheless, a bicycle fitted with an engine too small to be used for anything but a booster device would still be an extremely energy efficient vehicle. Furthermore, such a design requires absolutely no technological advancement—it is ready now.

Overcoming Weather Limitations. The only feasible way to overcome the problem posed by rain is to provide a complete or partial enclosure for the vehicle. Since this is very difficutl to do for a vehicle that is not self-standing, an enclosed tricycle is probably more practical than an enclosed bicycle. Such enclosure would also serve as a protection from wind burn and chill, though it would have to be carefully designed to minimize air resistance. In addition, the cold weather problem would be greatly mitigated by a proper enclosure design. If the enclosure were not excessively drafty, no external heater would be required for most cold weather conditions since the human body generates plenty of its own heat when exercising. As long as this heat is trapped, chill will be largely avoided as is illustrated by the effectiveness of very light-weight ski clothing in keeping the exercising wearer warm.

The enclosure for the three-wheeled pedal car must, of course, be as light as possible, as must the entire vehicle. Such a human-powered car could be designed for more than one passenger and could even provide for some freight carrying space. It would be absolutely critical to incorporate a highly efficient gearing system into the vehicle's transmission, not only because of its greater weight but also because the occupants would be constrained from standing up—a great help when pedalling uphill.

Several human-powered vehicles have already been designed along these lines. Some have even been offered for retail sale. Total weight has been held to under 100 pounds, in some cases, and such vehicles have been shown capable of negotiating inclines of up to 20 degrees.[31] One Canadian design includes two seats, baggage space, automobile-type headlights, multigear transmission, and dual pedal drive.[32]

Freight Carrying Bicycles. It is not at all difficult to carry cargoes of up to 20 pounds or more by bicycle. In fact, substantially more might be readily carried over favorable terrain for short distances by a cyclist in good condition. A bicycle or pedal car with an efficient gearing system could therefore be used for light to moderate shopping trips as well as for urban delivery of small amounts of freight. It makes little sense, for example, to expend the energy necessary to move a 3,000 pound, or even a 2,000 pound, motor vehicle, generate the associated pollution, and possibly block traffic by double parking at the delivery points, all in order to deliver a few ounces to a few pounds of medicine. A bicycle is perfectly capable of handling such urban freight with much lower energy expenditure, no pollution, no parking problems, and substantially reduced delivery time.

The same design changes that would help to overcome terrain and weather problems would also increase the effectiveness of bicycles as urban delivery vehicles and/or expand their usability for family business-type shopping trips. The only additional consideration is the design of the cargo compartment itself.

It must be as light as possible and ideally should be collapsible and detachable. Strong, light-weight, detachable nylon cargo bags are presently available.

Improving Bicycle Safety. The basic design of the men's style bicycle is strong and does not pose a danger problem in and of itself. Individuals with a balance problem, or with otherwise impaired reactions or coordination, could achieve the essential advantages of bicycling by using a tricycle. Such machines can be elegantly designed and styled and may provide an extremely attractive and inexpensive mobility increasing device for the elderly and for the handicapped.

The critical problems in bicycle safety are the dangers posed to the cyclist by motor vehicles and road defects (e.g., potholes) and the dangers posed to the pedestrian by the cyclist. And yet the solution to this problem could not be simpler. Providing bicycle paths usable only by human-powered vehicles, completely segregated from pedestrian sidewalks, on the one hand, and roads, on the other, will go far in keeping everyone out of everone else's way.

Along highways a portion of the road shoulder could be set aside for cyclists. In city streets, either part of the sidewalk, part of the roadway, or both can be established as exclusive bicycle paths. If space is very restricted, elimination of parking on one or both sides of the street might provide the required room. At any rate, a ban on incursions of pedestrians, bicyclists, and motorists on one another's right-of-way (except in emergencies and at appropriately designed and controlled intersections) should be strictly enforced. Such systems are currently in existence in various places throughout the world (for example, in Amsterdam) and they apparently function quite well. In most cases, the capital investment required to establish these segregated urban rights-of-way is either small or virtually nonexistent, and their establishment is critical to the safety and attractiveness of cycling on a wide scale.

Cycling and the Education Problem. Only in the past few years has the bicycle again come to be accepted as more than a toy in the United States. But its usage is still largely confined to purely recreational and exercise purposes. In this automobile-oriented society, people rarely think about using a bicycle (or walking), even for relatively short trips.

As we have discussed, the bicycle is not only an incredibly energy efficient mode of transportation but a completely nonpolluting one as well. It is cheap, it is heathful, and with proper rights-of-way, it is safe too. If all of the modifications described in the last section were adopted, the bicycle would be an even more attractive mode of transportation. And yet, unless people in this country come to understand that the bicycle is not merely a toy or an exercise machine and stop thinking of commuter bicyclists as adversaries encroaching on the motorists' territory, the substantial energy and ecology gains derivable from a maximal shift to bicycles will never come to pass.

With the possible exception of the establishment of special bicycle rights-of-way, an educational program aimed at explaining the speed advantage and the financial, energy, ecological, etc. benefits of bicycles may be the most important contribution toward inducing their widespread use.[j]

Buses

Although buses are very energy efficient, they are generally not sufficiently comfortable to be an attractive means of transportation for distances greater than 200 miles or so. Some design modification would go a long way toward improving their comfort and thus their acceptability, even for distances of as little as 50 miles. If buses traveling these distances were made about a foot wider and a few feet longer, without an increase in the number of seats (or kept the same size and fitted with a reduced number of seats), enough additional seat room would be available to make the buses significantly more comfortable. There would naturally be some sacrifice in energy efficiency, but to the extent the additional comfort attracted ridership away from the automobile, there would still be a significant net energy savings.

Urban buses in particular would benefit greatly from the replacement of their diesel engines with quieter, less polluting engines that did not have to idle when the buses stopped. In the last chapter, we saw that both the Rankine and Stirling engines possessed these important characteristics, and both of these engines have, in fact, been installed and operated in buses. The Lear Motors Corporation designed and built a steam turbine bust that was operated for six weeks on three different city bus routes in San Francisco—routes that included grades as steep as the steepest street in that city. The bus apparently performed effectively, with emissions only 36 percent as great as those allowed by the 1975 California standards, but its fuel economy was relatively poor. An improved version of the bus is claimed by the Lear Company to be presently competitive with the diesel in fuel economy over urban driving cycles.[33] The Philips Company of the Netherlands has likewise installed a 200 horsepower Stirling engine in a DAF bus, which has been demonstrated in the Netherlands for several years.[34]

Aside from enlarging the longer-range buses and replacing the diesel engines of, at least, urban buses, with more appropriate engine types, about the only other improvement in the buses themselves that could enhance their attractiveness would be the provision of easily accessible package storage spaces in urban buses.

The outstanding logistic advantage of the bus is its extreme route flexibility.

[j]Motorcycles have not been seriously considered in this section because they are more like light-weight cars than they are like bicycles. They are highly energy efficient but are thought to be simply too dangerous, except in very low-powered scooter or motorbike versions, to be a reasonable transportation alternative for general use.

If exploited properly, this characteristic may be used effectively in making the urban bus, in particular, a highly attractive, low capital investment, energy efficient means of mass transportation. But this is best discussed in the context of urban transportation systems as a whole, as in the following section.

Urban Transportation Systems

As of 1970 more than 95 percent of all urban passenger traffic was carried by the energy inefficient private automobile. The resultant congestion and pollution have contributed importantly to the decline of the quality of urban life in the U.S. The automobile certainly has its place in the urban transportation system, but automobiles and their associated roads and support facilities should *not* themselves *be* the urban transportation system.

To be effective, an urban transportation network should be developed within the context of the particular urban area in which it is to function. It should be tailored to the specific characteristics of the area and the needs of both the resident and transient population. It is all very well to start from scratch and design the ideal transportation system as an integrated part of the ideal metropolitan area. But we live in a world that is sadly much less than ideal, and if we wish to take full advantage of the opportunities for improving it, we must at least root our strategies in the existing reality.

The strategies discussed in this section are consistent with a considerable variety of particular urban transportation systems. If adopted, they would at the same time improve the quality of the transportation system and reduce its energy consumption.

The Commuting Network

There are three components to a commuter transportation system: collection, main route transportation, and distribution. *Collection* involves the movement of passengers from their individual trip origins to the start of a main route downtown. *Main route transportation* then carries them to a point in the downtown area from which the *distribution* system moves them to their individual destinations. The systems is, of course, reversible for the commute home.

For the most part, the same modes that can be used for collection can also be used for distribution, but the greater density of the downtown area does not generally make this approach optimal. The density differential would tend to imply a greater emphasis on mass transit, or at least public transit, downtown. For example, private automobiles may be an effective collection mode for a relatively dispersed population, but they are clearly inferior to buses, subways, or even taxis as a distribution mode.

Collection Modes. Buses, jitneys (i.e., vehicles that might either be considered small buses or large taxis), taxis, private cars, bicycles, and walking are all feasible collection modes. Walking is always the preferred mode for individuals living very close to the public main route transportation system, but this is unlikely to be any significant fraction of the population in a suburb. Bicycles (or pedal cars) on the other hand are an extremely attractive collection mode that should be usable by a considerable percentage of the commuters.

Aside from the obvious energy and pollution abatement benefits of human-powered vehicles, a mile or two of bicycle or pedal car riding in the morning and evening should do wonders for the vast majority of workers who hold essentially sedentary jobs. In addition, a ride of, say, 2 miles at an average speed of 8 miles per hour would only require 15 minutes and would scarcely raise a sweat where the terrain was congenial and the temperature less than excessive. Thus, even the typical dress codes of commuters are not inconsistent with this use of bicycles. Furthermore, commuters do not generally need to carry heavy cargo, so that freight weight should not really be a problem. However, the security of bicycles left at main route modes by commuters may be a problem in some areas, so that sturdy bicycle racks to which bicycles may be securely fashioned would be required.[k]

Private cars may be an excellent means for the collection of commuters living at fairly long distances from the main route nodes. These may be used in either the so-called part-and-ride or kiss-and-ride modes. In *park-and-ride,* the commuter drives to the main route node, parks the car, and takes public transit downtown; in *kiss-and-ride,* the commuter is driven to the station and dropped off by another individual who then drives away with the car. Though the former practice is more capital intensive, in effect, tying up a car for most of the day and requiring parking space, it is more energy efficient because the collection distance need only be traversed twice rather than four times each day.[l]

The capital intensity of park-and-ride can be reduced in two ways. Since the distance to be traveled is quite limited, the car is only used over that distance twice a day, and no significant cargo capacity is required, extremely small, lightweight and low-powered vehicles can be used. This will reduce capital intensity both directly and indirectly, by reducing parking space. It will also enhance energy efficiency considerably.

The park-and-ride system can be made more efficient and less expensive to establish by proper choice of park-ride lot locations. The best park-ride lot is one that is not only well located relative to a main route downtown, but also one that can be used for other purposes at night. For example, a lot located near a

---

[k]Motorbikes, motorcycles, and scooters might be acceptably efficient modes for more extended distances.

[l]Assuming that the person driving the "kiss-and-ride" car away does not have to pass by the station twice a day (at the right times) anyway.

main highway or train line could be used for park-ride during the day and for drive-in movies at night. Or the parking lot of an athletic stadium or racetrack could be used for park-ride, if agreements were made to schedule daytime events there only on weekends and holidays.[m] Alternatively, park-ride lots could be established in the areas next to (i.e., either inside or outside of) existing highway cloverleafs or at rest stops without heavy daytime use.[35]

Except in areas of unusually high density, buses are not an effective collection mode. In low-density areas, they must stop at too many places in order to pick up sufficient passengers for a reasonable utilization of their capacity. This extends collection time inordinately. Dedicated taxis are effectively paid kiss-and-ride collection (with due apologies to taxi drivers), which tends to be no more energy efficient and may be considerably more expensive.

Jitneys with a capacity of 9 to 15 passengers represent a compromise that may be quite energy efficient in areas of high moderate to moderate suburban density. Such vans have been used effectively for commuter collection on a door-to-main-route basis when scheduled in accordance with telephoned requests for service.[36] The critical conditions for the energy, performance, and financial efficiency for such "dial-a-ride" services are sufficient geographic population density and sufficient density of travel requests in time (ideal in commuting rush situations).

Main Route Transportation Modes. The primary alternative energy efficient main route modes are the bus and the train. The bus is advantageous in that its routing may be readily altered, while the train's main advantages are its higher speed and greater capacity flexibility.

If buses are to be used as main route transportation, special express bus lanes should be set aside on the commuter route highways, especially at bottlenecks, during rush hours. Automobiles, except possibly those being used in carpools, should be prohibited from entering these lanes. A bottleneck experiment of this sort was tried with some success on the San Francisco–Oakland Bay Bridge.[37]

One of the interesting ways in which bus route flexibility may be used to advantage is by retaining a portion of the express buses in the downtown area to serve as part of the city's internal mass transit system during the day. These same buses would provide express bus service in the evenings back to suburban locations where they would be available for nighttime, weekend, and holiday transportation service. In this way, the bus fleet would, in effect, be following the flow of population with a resultant substantial increase in the energy efficiency and quality of service of the transit system. In addition, the buses would be

---

[m] It does not matter which use is established first. The only important considerations are the suitability of the lot location and the sufficiency of its capacity.

utilized at a much higher percentage of capacity, making their cost per passenger of providing service correspondingly lower and thus allowing fares to be held down.

It may be possible to achieve a similar effect with trains, where the downtown area justifies the existence of a subway system. Commuter trains could be used in that system during the day. There is more likely to a vehicle compatibility problem here, however.

Whether trains or buses are used for main route transportation, one of the keys to maintaining high effective trip speed is a high degree of coordination with the collection and distribution modes and a small number of stops.

Distribution Modes. Walking is much more feasible for distribution than for collection, assuming a high-density city center. With the improvement in air quality and the reduction in noise that these more efficient urban transportation systems should bring, it might even be pleasant when the weather conditions are not too extreme. Bicycling would be efficient, but its feasibility is doubtful unless the problem of secure storage of bicycles overnite and on weekends near main route downtown modes is solved. Even so, the commuter would have to have a place to store the bicycle at work and would not have access to the bicycle for most noncommuting uses. Therefore walking seems a much more practical human-powered means of commuter transit than bicycling in the downtown areas.[n]

Buses and subways are excellent distribution modes, though the latter are so expensive as to require a very high level of downtown traffic to be cost effective. Taxis or jitneys can be reasonably efficient from a transportation point of view, if they are dispatched in an orderly fashion from special queues established at the downtown mainroute transportation modes. A controlled taxi dispatching experiment of this nature was tried at the busy Port Authority Bus Terminal in New York City, with a resulting reduction in waiting time, increase in average passengers per cab (and hence in energy efficiency), and substantial diminution of traffic congestion.[38] It is also common practice in Japan.

It has been estimated (for the year 1971) that an energy savings of nearly 50 million Btu's per year would result for each individual who shifted his or her work trips from automobile to bus.[39] This implies that if we develop an effective commuter network of the type discussed here, it is not unreasonable to expect energy savings of roughly the equivalent of 400 gallons of gasoline per year per individual commuter. Thus, if only 1 million commuters were shifted from the urban auto to attractive mass transit, total energy savings of between 0.3 and 0.5 billion gallons of gasoline per year is a reasonable "ball-park" estimate.

---

[n]A colleague at Columbia University has apparently had some personal success with roller skating.

Carpools

As is indicated in Table 5–1, in 1970 a fully loaded automobile was as energy efficient as a bus at average actual loading in 1970. However, a fully loaded bus was still more than twice as energy efficient as a fully loaded car. What this shows is that although carpooling renders automobiles much more energy efficient, commuting by automobile (where traffic density is high enough to justify buses for main route transportation) is still highly energy inefficient compared to a park-and-ride system. Nevertheless, where the commuter density is low, or for noncommuting uses, carpools offer considerable advantage over individual car travel.

The effective use of carpools requires a system for matching departure times, places of departure, and destinations among potential participants. If the system is for low-density commuting to a given place of employment, probably the most efficient way of organizing the carpool is through the employer. The coordination problem is minor, even for a sizeable employee workforce, and should be handleable with minimal fuss. The participation problem should be solvable through education as to the financial advantages, priority assignment of parking, and other inexpensive incentives. Where there is no single employer involved, coordination is probably best achieved by a government-provided carpool facilitation service, perhaps operated roughly along the lines of a computer-dating service—though not necessarily computerized.

As for noncommuting carpools, these might conceivably be facilitated by a central posting service or switchboard, say, at an apartment complex. But on the whole, noncommuting, non-special-event carpools are probably best left to informal arrangements among friends and relatives and should be encouraged as part of a general ethic of energy conservation.

When carpooling, it is often more energy efficient to meet at a central place where all but one car is parked than to have one driver pick up all passengers at their individual origins, unless of course the individual origins are very close to each other. Consider, for example a T-shaped carpooling situation in which each of two participants lives at one end of the horizontal part of the T, say, 5 miles from each other, and the destination lies at the far end of the vertical part of the T, which is, say, 10 miles long. In a round trip, the total distance traversed will be 35 miles if one driver picks up the other at home and returns him or her there and only 30 miles if they rally at a central point between them.

Moving Within the Central City

Considering the city itself as a transportation area, there are several modifications of the design and operation of the components of the transportation system that can reduce energy consumption and air and noise pollution while facilitating intracity movement. Some, like providing pedestrian overpasses or underpasses at large, busy intersections to facilitate walking, are rather simple. Others are a bit more complicated. We should not, however, take the lack of attractive public transportation in most U.S. cities as in any way inevitable or unalterable.

Subways. Subway systems can be a very efficient, fast, and even relatively comfortable and quiet means of mass transit. The London Underground, the Paris Metro, and the Toronto subway system are some of the more rational systems that do not assault the senses as does the New York City subway, for example. Subways should be well-integrated with other modes of transit, as they are in Toronto, a city that is recognized as a leader in the field of urban people moving.[40]

Probably the most important design modification in the subway trains themselves, as far as energy conservation is concerned, is the installation of a regenerative braking system. More than any other vehicle, a subway train is an acceleration machine. It spends little time cruising relative to the time it spends either speeding up or slowing down—and it is massive, implying that the accelerating/decelerating forces are very large.

Presently, subways are commonly halted by either friction braking, which not only dissipates energy as heat but also deforms the wheels, or by dynamic braking. In dynamic braking, the electric traction motors of the train are run as generators, converting the kinetic energy to electricity, which is then dumped into a resistor, i.e., dissipated. Energy is wasted in either case, but at least with dynamic braking, wheel wear and thus wheel noise and vibration is substantially reduced.

Regenerative braking through the use of flywheels or direct energy storage in efficient batteries would make this deceleration energy available for reacceleration, as has been pointed out with respect to other vehicles. However, the potential energy savings are so great in the special case of subways that it is even possible that conventional heavy steel flywheels will yield a considerable net energy savings. Such a steel flywheel system is presently under study for use in the New York City subway system.[41]

Taxis. The availability of an urban vehicle that provides transportation over a passenger-determined route at a passenger-determined time can be an important asset to a total urban people-movinb system. And if such a vehicle can be used serially by a number of passengers rather than being dedicated to one particular passenger, as is a private car, this flexible service can be provided with minimal traffic congesting and space pre-empting effects. Hence the advantages of the urban taxi system as one means of achieving the desired results.

In order to offer maximally effective service with minimum energy consumption, the following conditions should be met: the vehicles themselves should be as energy efficient as possible and should maximize the ratio of usable internal cargo space to total volume (without significant regard to aerodynamic profile), taxis should be available within as short a distance and time of the place and time at which their services are required as feasible, and the amount of time during which the taxis are traveling without cargo should be minimized.

Maximal vehicle energy efficiency implies using small cars, specially designed for carrying passengers. The design should allow for easy passenger entrance and exit and comfortable seating. Because it is very uncommon for a taxi to transport more than two passengers at a time, seating should be provided for only two or perhaps three adults. Though this will occasionally result in two taxis being required instead of one to transport a small group of passengers (with consequent additional energy consumption), it will far more frequently result in a reduction in unneeded vehicle weight, producing an important net energy savings. Furthermore, for the same reasons, very little baggage space should be provided inside the vehicle—perhaps only enough for two moderate sized suitcases. Supplemental baggage carrying capacity could be achieved through the use of light-weight, roof-mounted luggage racks.

The practice of taxi cruising, i.e., keeping the taxi in motion while searching for passengers, is highly wasteful of energy because of the large fraction of the mileage driven without passengers—estimated at an average of 45 percent by one small-scale study of the New York City area.[42] The practice also increases air pollution, traffic congestion (thus wasting fuel in other vehicles and increasing their contribution to air pollution), driver fatigue, and traffic accidents.[o] In addition, it is not even clear that cruising increases either the availability of taxis to customers or the profits (or even revenues) of the taxi owners—its main alleged advantages. A well-developed, efficiently dispatched system of radio-equipped taxis and hack stands with single queueing would probably increase the effectiveness (and profitability) of the taxi system. while reducing noise, air pollution, and traffic congestion and conserving energy.

One such system, described by Mastorakis, would have a central radio dispatch center taking telephone calls from the public and from dispatchers on foot (or in small booths) at the major hotels, bus terminals, theatres, etc. throughout the area.[43] The radio center would process the calls and assign them to taxis within each call's zone of eligibility—say, within half a mile radius from the present cab position. Taxis neither carrying passengers nor proceeding to an assigned call would be required to move to the nearest hack stand and join the queue. These stands would be established in large numbers throughout the city. Taxis waiting on a queue that was not actively boarding passengers lined up in

---

[o]In 1973, 5.0 to 5.5 times as great a percentage of New York City taxicabs were involved in traffic accidents than the percentage of total vehicles so involved. This held true both for fatal and nonfatal accidents. (See Basil Mastorakis, "Energy Reduction in the Operation of an Urban Taxicab System," p. 7).

significant numbers would have to shut off their engines. Some sort of small-scale plug-in heating system could be provided at these locations so that drivers would not become excessively cold sitting in an unheated cab in cold weather.[p]

If this sort of system were to be established and the relevant rules effectively enforced, the resultant energy savings would be considerable. For example, if the system were to succeed in achieving net energy savings equal to only two-thirds of the energy now consumed in empty cruising, some 25 to 30 million gallons of gasoline would be saved annually in New York City alone.[44]

As an alternative to taxi systems, there have been proposals to provide pick-up/drop-off very small, low-powered urban cars for temporary private use. This type of system would be much like present car rental operations but on a vastly expanded scale with greatly simplified and expedited rental procedures and very short-term rentals (e.g., 15 minutes). Operationally, this system might involve possession of a special key obtained by payment of a periodic fee. On the whole, the feasibility of this sort of scheme seems doubtful with the present context of high urban crime rates in the United States.

Personal Rapid Transit Systems

One of the most intriguing urban transportation ideas to begin to approach technical feasibility within the last few years is that of the personal rapid transit (PRT) system. There are several versions of such systems presently in one or another stage of development, including the West German CABIN TAXI, the French ARAMIS, the Japanese CVS, and a few U.S. PRT systems. Though there are significant differences among the various versions, they involve a number of common elements. Thus it is possible to describe a typical PRT system.

Small passenger cars, each with a capacity of up to four occupants, ride along an elevated concrete guideway. The cars are computer-controlled, and the guideways resemble monorail systems more than elevated train tracks. Passengers assemble at PRT stations, board an available car (the cars can be summoned if they are not on queue at the station), insert a special plastic "credit" card into the control panel, and push buttons indicating the destination desired. The signal travels to the central computer which charges their "account" for the appropriate amount, and the car proceeds completely automatically to the destination. The route is not under passenger control, nor is the speed–both of these parameters are determined by computers that minimize trip times over the system by considering all traffic conditions, destinations, etc. for the system as a whole. It

---

[p]A system similar to that employed in some parking lots in some sections of Canada that face extreme winter cold might be used. There, poles with electric outlets are provided so that a small supplemental heater built into the car may be plugged into the outlet to keep the engine warm enough to prevent it from freezing. For hack stands, the heater would be for the driver's compartment only.

is, however, possible to allow for a system of trip priorities that will favor certain trips at the expense of others if desired.

The idea is that a widespread network of such guideways could provide a PRT system that is competitive, in terms of convenience and privacy with the private car, while consuming less energy. Cost estimates for such systems have been very optimistic. For example, one very interesting system designed for the Los Angeles area (an area notorious for its lack of mass transit) has estimated operating expenses of about 3 to 5 cents per mile and one-way guideway capital expenses on the order of 3.0 to 4.5 million dollars per mile.[45] However, there is no hard evidence to date that such a widespread PRT system is even technically feasible, much less that the favorable cost estimates that have been made have any relation to reality. In fact, very small-scale simplified versions of these or similar automated people-moving systems have run into tremendous economic problems and have experienced considerable technical difficulties.

By way of illustration, a small PRT system built at Morgantown, West Virginia was initially to have 5.4 miles of guideway, 6 stations, and cost of $18 million. A revised version of the plan provided for half as many stations, and its construction costs ran up to some $60 million.[46]

Therefore, one must be at least somewhat skeptical of the favorable cost, performance, and energy conservation estimates that have been made without real construction and operating experience. On the whole, while PRT may play a significant role in mass transit for some specialized uses in the future, it does not currently seem to be a real transportation alternative. Most especially, its total energy implications are particularly unclear—even if a technically feasible operating system were developed. It is not, however, an idea without promise.

Summary and Conclusions

A number of land transportation alternatives to the relatively energy inefficient automobile and truck have been considered. Various modifications of the design, operation, extensivity, and coordination of these modes have been suggested, primarily with a view to increasing their attractiveness in order to induce an energy-saving shift of traffic in their favor.

Though the energy efficiency of the railroads has increased considerably over the past few decades, their share of traffic has declined substantially. To reverse this latter trend, an increase in the speed, reliability, coverage, and comfort of rail service is required. One way of achieving this is to reduce the nearly universal reliance of United States railroads on diesel-electric locomotives. These engines have certain design limitations that restrict their power and thus the speed and the trailing weight of the trains they haul. They have also experienced some real reliability problems, chiefly with their cooling systems. Pure electric locomotives are an interesting presently available alternative that apparently have significant performance, reliability, and even energy efficiency advantages but require a considerable capital investment. In the longer run, it may well be that other engine designs, such as the turbine, will prove to be more effective still. This is an important area for research and development.

Serious improvement in the reliability of rail service, absolutely critical to inducing a shift of traffic to this mode, simply cannot be achieved without upgrading the poor track and replacing the aging rolling stock that are the legacy of years of neglect. Passenger service can be fast and comfortable, as is well illustrated by the rail systems in a number of European nations and Japan, and there is absolutely no reason why such high quality service cannot be established in the United States. The requisite technology exists, and there is reason to believe that effective passenger service is economically viable. The speed and reliability of freight service, on the other hand, would be greatly increased if improved materials handling, scheduling, and routing systems were employed. Expanded use of containerization offers considerable potential for expediting handling, as does expanded use of computers for routing and scheduling. Finally, an increase in the size of the rail network is an important asset in attracting an increased share of traffic.

Bicycles are an extremely energy efficient mode of land transportation, with considerable health, pollution reduction, and financial benefits as well. They are probably usable to replace at least a few percent of urban automobile traffic, despite their weather and terrain limitations, with a consequent energy saving on the order of a billion gallons of gasoline per year. The substitutability of bicycles for urban automobiles would be further enhanced by the use of a booster device for hills (e.g., a very small supplemental motor or a flywheel) and the modification of bicycles to leg-powered, enclosed tricycles or pedal cars to overcome some of the weather limitations. Bicycle safety could be greatly enhanced by providing a network of bicycle paths segregated from both pedestrian and motor vehicles rights-of-way along the side of every road. Finally, an educational program aimed at overcoming the tendency of people in the United States to think of bicycles as purely recreational would be of real value.

Buses used for longer-range travel should be made wider and longer. The chief design modification suggested for urban buses, on the other hand, is the replacement of their diesel engines with something on the order of a Rankine or Stirling engine, thus reducing noise and air pollution and potentially increasing energy efficiency.

The pollution and congestion of urban areas can be substantially reduced at the same time that energy is conserved and movement is facilitated by establishing effective well-coordinated urban transportation systems. The specifics of these systems should be determined in accordance with the peculiarities and needs of each specific urban area. It is unlikely that any one plan is optimal for all areas. But some important generalized energy conserving ideas are broadly applicable.

For commuting, the park-and-ride system is one such idea. Here commuters travel by car, bicycle, or on foot to a station from which a train or bus transports them (with few stops) to the downtown station. They move from the downtown station to their destinations on foot or by bus, subway, or taxi.

Carpools are not nearly as energy efficient but are clearly far superior to single-person car travel. Even carpools would sometimes save additional energy if they used a sort of park-and-ride, central gathering place system.

Movements purely within downtown areas are most efficiently carried out by bicycling or walking. However, when powered transport is required, subways, buses, and taxis are reasonable alternatives. Subways are energy efficient when fully utilized and can be fast, quiet, and comfortable. The fact that they are nearly always either accelerating or decelerating indicates that their energy efficiency could be increased substantially be employing flywheels (or similar devices) in a regenerative braking scheme. The utilization of buses can be enhanced by allowing a portion of the fleet to travel with the flow of traffic, operating in the central city during business hours and moving to the suburbs at other times. The energy efficiency of taxi service can be increased considerably by elimination cruising and replacing it with a system of hack stands and radio dispatching and control.

Automated personal rapid transit systems with small computer-directed cars operating on guideways have been suggested as an effective way of attracting travelers away from private automobiles. The systems, when fully developed, are supposed to provide the privacy and route/schedule flexibility of private cars at twice the energy efficiency of present cars in urban areas and at a relatively low price. But to date, the claims of economic attractiveness and efficient performance made for these systems remain unsupported, and in fact, what meager evidence does exist indicates that even simple, small-scale systems are extremely costly and not overly effective. The energy efficiency of urban travel can be increased at least as much as is claimed for personal rapid transit systems by following the recommendations of the previous chapter for automobile modifications or by shifting urban traffic to buses or subways. Thus, while these so-called PRT systems may prove attractive at some future date for some urban applications, they do not appear to be a cost-effective, energy efficient urban transportation alternative today.

In sum, more energy efficient modes of land transportation *can* be rendered attractive enough to draw an increased share of both freight and passenger traffic. The potential energy savings from inducing such a shift are very large indeed and could be achieved along with a substantial reduction in motor vehicle, air, and noise pollution without sacrifice in the material standard of living.

8 Fluid-Borne Transportation

It is estimated that there are 316,800,000 cubic miles of water in the rivers, lakes, and oceans of the world (only 2.7 percent of which is fresh)[1] Some 71 percent of the total surface of the earth is covered by the sea, and all of it is surrounded by the atmosphere.[2] Thus, water and air routes have a coverage and flexibility far greater than those restricted to the land. Furthermore, those routes are always continuous in the case of air travel and much more frequently continuous in the case of water than land travel. Air routes are also nearly always more direct than those restricted to the land surface.

For these reasons and others, fluid-borne modes of transportation are, and always have been, important to the efficient movement of passengers and freight. Our purpose here is to consider how the energy efficiency of such means of transportation may be increased, and how the services of the more energy efficient modes can be more fully utilized.

Differences between Fluid-Borne and Land-Based Transportation

An object sliding on the surface of a fluid does not, in general, encounter frictional forces that are as severe as those resulting from similar motion on a solid surface. However, to the extent that the object is submerged, its motion will be opposed by the drag effects of the fluid. All other things being equal, this fluid friction is greater the more viscous (i.e., the thicker) the fluid. Thus water produces far more resistance to motion, *ceteris paribus,* than does air, and hence far more energy is required to move a given object through water at a given speed than through air.

Fluids also exert forces that tend to counteract the pull of gravity, pushing the object upward. If this lift occurs purely because the object is lighter than the volume of fluid it displaces, it is called *static lift* (or buoyancy). If it occurs because the fluid is flowing faster over the upper surface of the object and thus exerting higher pressure underneath the object (the Bernouilli effect), it is called *dynamic lift.* Because static lift does not require any movement of the object relative to the fluid but rather is a property of their relative densities alone, it is possible to achieve static lift without the energy expenditure typically required to produce the motion necessary for dynamic lift.[a]

As a result of natural (thermal, gravitational, etc.) effects, the hydrosphere and the atmosphere experience considerable internal motion. These air and water currents tend to carry objects along, implying the need for less non-natural energy expenditure to transport those objects in the direction of the currents. Of course, for the same reason increased energy expenditure is required to move against the currents. Yet these effects do not necessarily cancel out, because there are often ocean or air currents flowing in both directions, though in separate locations. It is sometimes possible to exploit this situation, always traveling with the current and hence conserving energy.

Lower friction, higher drag, static and dynamic lift, and the existence of currents are all important differences between land-based and fluid-borne transportation.

Inland Water Transportation

Something on the order of two-thirds of the earth's land surface is drained by streams and rivers that flow eventually into the sea. On the one hand, these waters frequently meander, increasing the effective trip distance between any two points and thus requiring more energy expenditure. On the other, they flow continually downward from their source to the sea, avoiding the up-and-down hill problem common to land routes. Even with regenerative braking, the energy required for traveling uphill is never fully recovered during the trip downhill. Therefore the monotonic rise or fall of rivers and streams tends to be energy conservative.

Of course, not all parts of all streams and rivers are navigable. They are often too shallow and sometimes drop precipitously. But where they are usable (or can be made so by a relatively small investment), they provide an interesting and effective transportation "roadbed." Though rapidly increasing hydrodynamic drag results in a considerable energy penalty unless movement is relatively slow, the water tends to distribute the load thus reducing the stress on the vehicle itself. This, combined with low friction, implies that watercraft operating on navigable inland waterways are an efficient, low-speed means of transporting heavy or bulky cargoes. Fragile cargoes likewise benefit from the extreme smoothness that characterizes most navigable inland waterways most of the time. Water transportation can also be an efficient way of moving people where distances are short and water routes are more direct than those over land.

[a]Dynamic lift does not necessarily require movement of the object relative to the surface of the earth but only relative to the fluid. If the fluid has already been placed in motion by natural forces, the object may be lifted dynamically without its having to expend any additional energy, e.g., a kite flying in the wind.

Under these circumstances, the effective trip speed may be greater for water transportation even though the vehicle speed is considerably lower.

In the United States as of 1970 about 45 percent of the population and 80 percent of the metropolitan areas were located on or near one or more bodies of water.[3] Therefore, it is not only possible to move goods and people to and from a sizeable fraction of the nation's population and urban areas by water, it is also possible to integrate water transit into the internal transportation network of most United States cities. In too many places, we have come to regard bodies of water as obstacles to travel to be bridged over or tunneled under. It is time to more fully integrate water transportation into our daily short-distance transportation system. The energy advantages of doing so should be considerable, since watercraft have about the same energy efficiency as railroads (see Table 5-1).

The slowness of water transportation, often considered to be its chief liability, can actually be an asset for some classes of cargo. In particular, as previously discussed, any cargo that is merely being transported to a storage facility receives free in-transit storage during the period of its movement, often producing a net reduction in total storage costs. Similarly people may sometimes be "stored" overnight for a lower fee on-board a ship than at the hotel to which they are traveling.

The effectiveness of the inland water transportation system can be increased by using canals and/or canal-lock systems to either connect disconnected but proximal bodies of water or to overcome the navigability problems created by rapidly dropping waterways. At one time such systems were common in the United States. In fact, in the quarter century following 1815, some 3,326 miles of canals were constructed.[4] Probably the most famous of these was the Erie Canal, in large part responsible for the rapid growth of New York City during that era. However, by the mid-nineteenth century the canals were competed out of business by the railroads. While it may not be desirable to rebuild a full-scale canal system, it is worth considering the possibility that restoring and modernizing at least parts of the system may have some energy conservation value.

Sea Transportation

In considering sea-going vessels, attention will be concentrated primarily on freight uses. Given the extreme disparity between the speed capabilities of ships and aircraft (e.g., tran-Atlantic crossing times of 4 days versus 6 hours), it seems highly unlikely that sea-going vessels will recapture any significant segment of the passenger market within the foreseeable future. Increased speeds are conceivable with designs such as the hydrofoil, which utilizes the dynamic lift provided by wings set into the water in order to raise the vessel largely above the water's surface, thus drastically reducing hydrodynamic drag. But these designs have been far less energy efficient than standard vessel configurations and are still not competitive with aircraft in speed. Therefore, although some of the discussion will be applicable to shipping in general, the passenger market will be largely conceded to aircraft, at least, for transoceanic travel.

Power Systems[5]

Ships typically have two power systems: a main propulsion system, which drives the ship through the water; and auxiliary engines, which run the ships navigation instruments, cargo handling devices, lighting, heating, etc. Low- or medium-speed diesels or steam or gas turbines are widely used as main propulsion power plants, while the auxiliary function is most commonly performed by medium-speed diesels coupled to electric generators. Despite the fact that the ships' main propulsion systems often run on lower-grade petroleum distillates (called Bunker C or Fuel Oil) than do auxiliary engines, they consume several times more energy and therefore offer the largest a priori potential for energy conservation.

Low-speed diesels (with engine speeds on the order of 100 rpm) can be coupled directly to the propeller without the use of heavy, bulky reduction gears. They can function on low-grade, high-viscosity oils, provided the oils are preheated and filtered. The resulting sludge, however, produces a disposal problem.[b] Medium-speed diesels (500 to 1,000 rpm), on the other hand, have lower component weights and some maintenance advantages, though they do require reduction gears. They also have low headroom requirements, which is an advantage in certain types of ship designs.

Steam turbines are powerful, reliable engines capable of easily coping with the dramatic increases in ship sizes and power requirements of recent years. Though present designs are not nearly as energy efficient as diesels, their maintenance requirements are far lower. Thus the dominance of labor costs over fuel costs combined with the higher degree of capital utilization possible with increased ship availability (i.e., reduced breakdowns) have militated against the diesels and in favor of the steam turbines. Gas turbines are even less energy efficient than steam turbines but have gained some favor because they can be run with reduced crews.

A random sample of the fuel consumption rates of 40 ships, 10 in each engine class, gives a rough indication of the relative fuel efficiencies of these different engines. The sample, drawn from trade journal published specifications, shows that, on the average, medium-speed diesels, steam turbines, and

[b]It is too ecologically damaging to dump the sludge overboard. The best approach may be to break down the sludge into a form that can be burned in the engine.

gas turbines consume about 15, 45, and 54 percent, respectively, more fuel per hour (adjusted for engine horsepower) than low-speed diesels.[6] Thus, even if our attention is restricted to present engine designs, it is clear that it is technically possible to substantially reduce fuel consumption by opting for low- or at least medium-speed diesels. This should not restrict ship size, because diesels are capable of powering even the largest ships.

Fuel Consumption Parameters

As is true of every mode of transportation, speed and weight (mass) are two of the most important energy consumption parameters. Since we are dealing with freight transportation, cargo weight is a prominent component of total weight and, with some exceptions, best dealt with elsewhere,[c] reduction in cargo weight is self-defeating. However, large ocean-going vessels carry hundreds of tons of fuel. Hence, any increase in energy efficiency will produce a corresponding reduction in fuel requirements and thus a reduction in weight, which will further increase energy efficiency. Of course, a switch to lighter-weight fuels would mitigate the weight problem too, but the depletion of fuel resources would tend to be aggravated by the need to refine more crude to get these higher-grade oils and would not utilize existing fuel supplies as fully.

According to a paper by Heinen, Schane, and Bielefeld, which models fuel consumption as a function of velocity and fuel remaining in the ship's tanks, fuel consumption is minimized if a ship is operated at a constant speed throughout its entire voyage.[7] This conclusion is not hard to take because constant-speed operation avoids most energy losses from repeated decelerations and further makes possible the constant operation of the ship's propulsion engines at optimum design speed.

As far as absolute speed is concerned, the energy savings available from speed reductions may be relatively small for low speeds but is dramatically greater at higher speeds, because of the rising hydrodynamic drag effect. For example, reducing ship speed by 33 percent from 12 knots to 8 knots would produce about a 25 percent reduction in fuel consumption, while reducing the speed another 25 percent to 6 knots would only result in about a 2 percent additional energy savings.[8] As a further illustration, a comparison of two cargo vessels with the same cargo capacity, one powered by a low-speed diesel at 14.5 knots and the other powered by a gas turbine at 26.5 knots, reveals that the faster ship consumes almost 375 percent more fuel per mile, though its speed is less than 85 percent greater.[9]

Hull design and operating condition is another important factor in the

[c]The most notable exception is the reduction in gross cargo weight per unit achievable by reversing the trend toward overpackaging, a practice that is resource depleting, pollution generating, and energy wasting both in manufacture and in transportation.

ship's energy consumption. Over time, the roughness of the wetted surface of the ship increases as the paint deteriorates, the hull plates corrode, and algae, barnacles, and other encrusting organisms grow. Although anticorrosive paints and other protection devices can keep hulls free from corrosion for 18 to 24 months under most conditions, organism fouling typically begins within 6 months after drydocking. The increased hull roughness results in about 10 percent more fuel consumption on the average, by one estimate.[10]

Improving the Energy Efficiency and Attractiveness of Sea Transportation

Ships moving through the water, like vehicles moving through the air, will encounter less fluid resistance to their motion if they have better fluid-dynamic profiles, i.e., if they are more streamlined. According to Telfer's analysis of fuel consumption versus speed, a nonstreamlined ship traveling at 8 knots, for example, would consume 10 to 15 percent more fuel than a similar streamlined vessel.[11]

Improved hull maintenance for the purpose of reducing the average roughness of the wetted hull surface should be able to reduce fuel consumption by roughly 10 percent relative to present levels.[12] If this estimate is correct, it would mean that something on the order of 10 million tons of fuel oil could be saved each year by more frequent hull cleanings and better anticorrosion systems. Recent developments in the area of automated underwater hull scraping systems are therefore most encouraging, since they would allow an increased frequency of hull cleanings without requiring increased drydocking, an expensive and time-consuming procedure. At least one such system, relatively small and simple in concept, is already in operation.[d]

As for speed, it seems highly unlikely that the enormous increases in energy consumption associated with increased vessel operating speeds are justifiable. Rather, there are two far more energy efficient ways of increasing effective trip speed while maintaining or reducing energy consumption per ton-mile. The first is by continued increase in average cargo capacity per vessel. A 100,000 ton supertanker, for example, is even more energy efficient than a large pipeline and nearly an order of magnitude more efficient than a 15,000 ton containership.[13] Larger ships traveling at similar speeds deliver more cargo per unit of trip time, and thus, in a sense, increase the rate of flow of goods, i.e., the effective trip speed.

Effective trip speed may also be increased, as in the case of the railroads,

---

[d]This system consists of a 6-foot diameter disc that reportedly adheres to the hull with a clamping force on the order of 1,000 pounds. The disc contains scraping brushes and may be either diver-controlled or controlled from a remote console. It moves along the hull cleaning it underwater while the ship is at rest.

by improved materials handling techniques and improved coordination with other modes. Increased use of containerization should help to reduce materials handling delays while facilitating intermodal connections. For example, the trailer of a truck, the body of a railroad car (minus the wheels and undercarriage), or even an entire inland water barge may be loaded on a large sea-going cargo ship as a single container and subsequently unloaded as such at the point of destination. Of course, the proper design of the containers (especially with respect to minimizing their weight) is critical to the energy and time savings available, but the design requirements are simple and easily achieved.

Aircraft

During the same period in which the energy efficiency of the railroads increased by nearly a factor of five, the energy efficiency of aircraft declined dramatically. The DC-3, the typical commercial aircraft of the 1940s, consumed 2,630 Btu seat-mile, while the DC-8 (and the Boeing 707), the typical jetliner of the 1960s, consumed about 4,000 Btu/seat-mile.[14] But, whereas the DC-3 cruised at 150 mph, the DC-8s cruising speed was 525 mph. Thus aircraft had achieved a 250 percent increase in speed at the expense of only about a 50 percent increase in energy consumption. Since, as we have seen, transportation energy requirements tend to increase more rapidly than speed, the fact that aircraft speed was more than tripled while fuel consumption increased by only half-again as much is a reflection of the enormous advances in the state of the art during those few decades. Nevertheless, the energy effficiency of aircraft did decline substantially and air traffic did expand rapidly, and the combination of these trends contributed to the overall decline in transportation energy efficiency.

In the 1970s there were two important developments in aircraft technology, with exactly opposite energy consequences. The first, the development of the "jumbo jet," resulted in a tremendous improvement in the energy efficiency of air travel, reducing Btu consumption per seat-mile to 2,700 (less than 3 percent greater than the levels of the 1940s), while increasing speed modestly to 575 mph. The second, the development of the supersonic transport (SST), resulted in speeds on the order of 1,200 mph and energy consumption projected to be between 55 and 100 percent greater than that of the DC-8. The choice could not be more clear. We can travel from New York to London at essentially the same speed with 33 percent less energy consumption or make the journey twice as fast while burning up between 50 to 100 percent more fuel. Apart from any other considerations, it seems incredible that saving a few hours of flight time could conceivably be worth the energy cost.

The energy efficiencies of a number of different classes of aircraft are given in Table 8-1, in terms of gallons per passenger-mile. The efficiencies are esti-

Table 8–1
Estimated Energy Efficiencies of Selected Aircraft: Actual Loading[a]

Aircraft	*Passenger Loading*	*Passenger-Miles per Gallon*
Helicopter	24	5
Supersonic Transport	150	11
Corporate Jet	8	11
Tilt-Wing VTOL[b]	48	18
Standard Airline Jet (DC-8, 707)	80	20
Jumbo Jet (747)	200	30
Jet Airbus (DC-10)	180	33
Private Plane	3	37

[a]Estimated actual loading in the case of operational systems, projected loading in the case of the SST and VTOL.
[b]VTOL = Vertical Take-Off and Landing craft
Source: adapted from Richard A. Rice, "System Energy and Future Transportation," *Technology Review*, January, 1972, p. 34.

mated in accordance with actual (or projected) capacity utilization rather than at full loading. As Table 8–1 indicates, the present generation of airline jets is fairly energy efficient in comparison with other aircraft. This is particularly true of the larger jets, even though their assumed capacity utilization is only between 55 and 75 percent, i.e., they typically flew (in the early 1970s) at least one-quarter empty. In fact, their passenger-miles per gallon are roughly comparable to a Volkswagen beetle carrying the driver alone.

Table 8–2 focuses more specifically on the jet aircraft in common use by the larger air carriers. Fuel consumption rates are given as a function of seating capacity (in standard configurations) rather than on a per passenger basis. If we ignore, for the moment, the modified extra capacity versions (given in parentheses) and consider only the basic models, we see the following pattern: the intermediate- and long-range standard jets consume fuel on an average of about 25 percent faster than do the short-range jets, and the short-range jets consume fuel about 15 percent faster than the long-range wide jets. This indicates that the basic energy disadvantages of greater total aircraft weight can be overcome if the planes are designed to maximize seating capacity. Comparing the fuel consumption rates of the enlarged capacity and standard versions of the DC-9, 727, 707, and DC-8 further strengthens this point.

Increasing Airline Energy Efficiency

It is clear that the enormous comparative advantage of high-speed aircraft as a mode of long-distance passenger travel is not likely to be seriously challenged in the foreseeable future. Since we have consistently required that energy con-

Table 8–2
Approximate Fuel Consumption Rates of Common U.S. Airline Jets

Plane[a]	*Number of Seats*[b]	*Fuel Consumption Rate (gal/seat/hr)*
Short Range		
DC-9–10	67	12.70
(DC-9–30)	(89)	(9.56)
737	92	9.74
Intermediate Range		
727–100	96	13.99
(727–200)	(127)	(11.52)
Long-Range Standard Body		
707–100	128	13.99
(707–300B)	(142)	(12.61)
DC-8–51	127	14.10
(DC-8–61)	(185)	(10.89)
Long-Range Wide Body		
DC-10	248	8.73
L1011	248	10.06
747	348	10.72

[a]Planes whose designation and data are in parentheses are enlarged capacity versions of the same models.
[b]Standard Civil Aeronautics Board configurations.
Source: Air Transportation Association of America, *Fuel Capacity and Fuel Consumption Rates–Airline Jets* (Washington: ATA, October 30, 1973).

servation be carried out within the context of maintaining or improving the standard of living, we accept the continued operation of high-speed passenger aircraft as given and consider what may be done to save energy in the day-to-day operation of public air passenger service.

Increase Unit Aircraft Capacity. As we have seen in Tables 8–1 and 8–2, aircraft designed to maximize passenger-carrying capacity tend to be more energy efficient. This has two important implications: first, the airlines should turn increasingly toward air buses and jumbo jets; and second, first-class seating should be eliminated (or at least severely restricted). There are now a number of large-capacity jets, ranging from the two-engine European A-300B Airbus to the three-engine L1011 and DC-10 and, finally, the four-engine 747. The A-300B, the newest and smallest of the four, can carry more than 260 all coach seats, while the DC-10 and L1011 can carry well over 300, and the 747 well over 400 in a similar configuration.[15] By moving to all coach seating, the capacities of these planes can be increased by 20 to 25 percent over present standard first-class–coach configurations. Similar relative gains are achievable by eliminating first class on other smaller planes.

If for one reason or another it is considered necessary to offer differentiated classes of service, this can be done without the energy penalties currently associated with luxury service. A section of the cabin can be set aside where superior food and drink is served, where there is a larger selection of entertainment, where there are more comfortable seats, more elegant decorations, etc. As long as the space occupied per seat is essentially equivalent to what is now called coach service, there will be no serious energy penalty.

Increase Capacity Utilization. It clearly makes no sense to increase unit aircraft capacity, or even maintain it, unless high levels of capacity utilization can be achieved. Empty seats represent excess weight with no purpose, resulting in a net energy penalty. In fact, it is not an exaggeration to say that attaining persistent high levels of capacity utilization is the single most important factor in improving the energy efficiency of air passenger travel.

Of course, if this were achieved by attracting sufficient additional traffic to the airlines to fill up existing flights, total transportation energy requirements would rise not fall, although the airlines' passenger-miles per gallon would decline. Whether this traffic was diverted from other more efficient modes or newly generated, the energy consumed for travel would still increase. But we have consistently pressed for changes that would make other modes more attractive, and hence divert traffic from the airlines. Therefore, it should be understood that the only legitimately energy conservative increases in aircraft capacity utilization are those that are achieved within the context of a decline in at least relative, and hopefully absolute, air traffic.

Presumably, one could increase capacity utilization by simply not allowing any airplane to take off until it had achieved some preestablished load factor, but it does not seem possible to run a rational air transportation system in this way. A far more practical and nearly as effective technique is simply to reduce the number of flights scheduled. If this is done in a coordinated fashion for the airline system as a whole, the inconvenience caused to travelers by reduced flexibility in the choice of departure/arrival times can be minimized. In most cases, the maximum displacement in flight times will probably be from a few hours to half a day, and the average displacement far less. The airlines have already cut back schedules enough to produce some increase in load factor under pressure of the energy crisis, but more substantial adjustments may be required before capacity utilization is as high as it could be.

Eliminate Short-Distance Flights. During the intial phase of a typical flight an enormous amount of energy must be transferred to an airliner to increase its potential and kinetic energy to those levels consistent with its cruising altitude and speed. Similarly, this energy must be slowly removed from the plane to bring it safely to a halt on the ground. During landing, the potential energy is readily removed by simply reducing the energy expended to counter the pull of gravity, and thus the removal of potential energy itself does not increase fuel consumption. But the safe removal of kinetic energy typically requires reverse thrusting of the engines and hence does consume additional fuel. Even if the engines of an aircraft were perfectly efficient, the fuel consumption rate during takeoff (and to a much lesser extent during landing) would be greater than that for level flight at cruising speed and optimum altitude. For example, a 747 can fly for about one hour at cruising speed with the fuel if consumes to take off and climb to cruising altitude.[16]

As a result, short flights will tend to be far more energy intensive than long flights. By one estimate, fuel consumption for trips under 200 miles is twice as great per seat-mile than for trips in the range of 600 to 1,000 miles.[17] Since these shorter distances are also in the range of trip lengths for which even standard trains and certainly high-speed trains are most competitive, it would probably be best to phase out all 200 to 300 mile flights entirely. In order to minimize the negative impacts that eliminating these flights would have on the performance of the transportation network, it is important that the rail system be improved sufficiently to provide fast, reliable service along at least the heavily trafficked short routes (e.g., New York–Boston, New York–Washington).

Drastically Reduce Air Cargo. Even a large cargo jet is a very energy intensive mode of freight transportation. A jumbo jet burns more than three times as much fuel per ton-mile as a 40-ton truck and more than 6 times as much as a 40-car freight train and yet is by far the most energy efficient of present-day modes of air freight.[18] The energy cost of air cargo is too high to be tolerable for all but the highest priority emergency shipments of medicines, highly valuable, light-weight and spoilable goods, and the like.

To the extent that the excess cargo capacity of existing passenger airliners is sufficient to cope with a significant fraction of the demand for this extreme priority freight, it may be carried with fairly small incremental energy cost. Unless this incrementally cheap cargo capacity is sufficient to cope with this demand plus the demand for air mail, it may be wise to reserve air shipment of mail for distances of say more than 1,000 miles. It should not be difficult to develop a rate structure or other regulatory system that would achieve these effects.

In order that these severe restrictions on air cargo do not produce a significant reduction in the standard of living owing to seriously lowered performance of the transportation net, it will be necessary to ensure that the displaced cargo will be able to be rapidly and reliably carried by alternative means. An effective railroad system should easily be able to perform this function. even at today's low speeds. A freight train averaging 30 mph can still carry cargo more than 720 miles per day if operated continuously and nearly 500 miles if operated 16 hours per day. The figures become 960 and 640 miles per day respectively for a "high-speed" 40-mph train.

Because of the railroads' advantage in being able to bring cargo from the heart of a producer area to the heart of a consumer area, a modernized well-run railroad system should be able to play the major role in the coordinated door-to-door delivery of freight over a 1,000 mile distance with an effective trip time of only a couple of days. There is very little freight for which faster delivery can truly be said to be crucial. This again highlights the importance of an efficient railroad system to the achievement of energy conservation without serious penalty.

Aerostats, in one form or another, may well prove to be an energy efficient alternative to present air freight for cargos that do require lower trip times than rail service can efficiently provide. This possibility is discussed in the next section.

Reduce Speed. Reducing the cruising speed of a standard airline jet from mach 0.84 to mach 0.80,[e] a 5 percent reduction, saves more than 10 percent of the trip fuel required for a 5,000 mile flight, for example, while only increasing flight time by about 20 minutes.[19] As speed is reduced further, the energy reductions that result tend to be progressively smaller. Under pressure of the energy crisis, the commercial airlines have already reduced cruising speeds somewhat, and it may be that the reasonably small reductions in operating speed that would result in significant energy savings without lengthening travel time unduly may have already been made.

Large gains in speed at the expense of large increases in energy requirements, as in the present supersonic transport designs, do not make ecological, economic, or energy sense. Civilian aircraft designers should instead turn their attention to increasing energy efficiency while maintaining speed. Our mania for increasing speed must be tamed, at least to the extent that the environmental and energy consequences of faster transportation are given heavy weight.

Whenever anyone advocates a position that sounds even remotely like saying "we can travel as fast as we ought to now, there is no use pushing it further," he or she is open to the accusation of stifling progress. And it is true, after all, that when the horseless carriage reached the incredible speeds of 20 and 30 miles per hour, there were those who considered them ungodly. But it is also true that the world is a finite size, so gains in speed result in decreasingly important reductions in travel time. If these progressive incremental reductions in travel time are achievable only at the expense of increasingly large increments in energy expenditure and ecological damage, there must be an optimum point at which the cost of achieving the next gain in speed is greater than the benefits

---

[e]The mach number is the ratio of the velocity of a moving body to the local velocity of sound. Mach 1.0 therefore indicates that the object is traveling at the local speed of sound, but since the speed of sound in air depends on air temperature and density, mach 1.0 represents different speeds at different altitudes. For example, at sea level at 15°C, mach 1.0 equals 761 mph, while at stratospheric heights (above 36,000 feet) it is equivalent to about 660 mph.

derived. Yet, even so, we are not arguing that travel speeds should not be increased but only that increasing air travel speeds *at the present level of aircraft technology* is simply too energy intensive to be worthwhile.

Optimize Altitudes. Since gravitational potential energy depends on height, the greater the altitude to which an airplane climbs, the greater the energy that must be expended to reach and maintain that altitude. However, the greater the altitude at which an airplane cruises (within the limits of its design), the lower the air resistance it encounters and hence the less energy is required to maintain speed. Therefore, for any given flight, there is an optimum altitude that minimizes total energy consumption. For short flights it will be lower, for long flights higher.

By way of illustration of the considerable impact of altitude on fuel consumption, both larger and standard jet airliners (e.g., 747, DC-10, and DC-8) consume on the order of 50 percent more fuel per hour cruising at 25,000 feet than they do at 39,000 feet.[20] As might be expected, the effect is somewhat smaller for the smaller airliners (e.g., 737 and 727), even apart from their lower ceiling altitude. For instance, a 747 consumes fuel about 40 percent faster at 25,000 feet than at 35,000 feet, while the difference in fuel consumption rates between those cruising altitudes is only about 30 percent for a 737. Nevertheless, the effect is still large.

Since we have already recommended that short-distance flights be essentially eliminated, the implication is that average cruising altitude for minimum fuel consumption should increase. Of course, safety considerations may sometimes make cruising at energy optimum altitude impossible, but where the safety factor is permissive, aircraft should be operated at energy optimum altitudes.

Minimize Holding Delays. Aircraft are often subject to delay at various points in the flight subsequent to the starting of engines. They may be delayed on the ground while taxiing out toward the runway (possibly waiting in line for permission to take-off), required to circle at various points enroute, or forced to enter holding patterns over destination airports awaiting permission to land. These delays all contribute to energy consumption and air pollution, as well as lengthening the trip time and disrupting schedules.

Improved flight scheduling and interairport communication should help to reduce these delays. Since congested peak-hour schedules and certain heavily traveled air corridors are responsible for a disproportionate number of delays, it may be wise to put an absolute ceiling on the number of take offs and landings permitted at any given airport and on the number of flights permitted along any given corridor per unit time. These limits should, of course, be determined with reference to the capacities of the facilities and routes involved to handle traffic without delay.

This ceiling limitation would also strengthen the tendency to make use of the more efficient large jets in order to accomodate the demand for air travel with fewer flights. Fewer flights would also make it easier for flight plans to be made with optimum cruising altitudes (less stacking problems) and reduce the strain on air traffic controllers, thus increasing safety.

When delays are unavoidable, they should be absorbed to the maximum extent possible by increased waiting time on the ground before the engines are started. When airborne holding delays occur, the planes should be held at the highest possible altitudes.

By one estimate, eliminating only the fuel-burning delays of 30 minutes or more (ignoring shorter delays) at the three major New York City area airports (La Guardia, Kennedy, and Newark) alone would save nearly more than 7.75 million gallons of jet fuel annually.[21] Lack of public data makes it difficult to estimate what eliminating shorter delays would save. Even though the total percentage of fuel burned by airlines in delays is small, it is not small in absolute magnitude. Furthermore, it is pure waste.

Adjust Taxiing Procedures. Aside from seeking to reduce taxiing delays, the taxi procedure could be rendered more energy efficient by either using trucks to tow the planes for most of the taxi distance or at least by using fewer engines while taxiing.

Aerostats

Some 40 years ago huge airships, some over 800 feet long, sailed through the skies equipped with lounges, dining rooms, and accomodations for 100 passengers. The German airship line founded by Count Ferdinand von Zeppelin carried 40,000 passengers during 4,000 flights and logged nearly 25,000 hours of flight time.[22] About 130 steerable zeppelin airships were built in Friedrichschafen, Germany between the turn of the century and the start of World War II, the first of which predated the Wright brothers flight by more than 3 years.[23] The subsequent rapid development of heavier-than-air craft, with their considerable speed and maneuverability advantages, soon put an end to the airship era. But the energy efficiency of the airship is so great, that there may be some real value in redeveloping this class of aircraft.

The basic energy advantage of the airship derives from the fact that it uses aerostatic rather than aerodynamic lift. Thus the airship need only expend energy for motion and not for either climbing or maintaining altitude. Despite the fact that these large ships encounter considerably increased air resistance, the reduced energy requirements resulting from the use of aerostatic lift are so great as to make the airship 5 to 10 times more energy efficient than the powered aerodynamic airplane.[24] The air resistance problem does, under present technology, place rather severe limits on operating speed: top speed may be only

130 mph,[25] while the most efficient cruise speeds can be anywhere from 10 to 80 mph lower than that, depending on design.[26] Of course, higher speeds can be attained but only at the expense of exponentially increased energy expenditure.

There are three main classes of steerable aerostatic airship, or dirigible. The simplest is the blimp, an airship without internal support or structure. It is, in effect, a gas bag whose hull shape is maintained by internal gas pressure. Then, there is the semirigid airship, which has a structural keel attached to the gas bag, but is otherwise quite similar to the blimp. Finally, the zeppelin has a rigid structural exoskeleton. The gas is contained in cells inside the rigid framework, and the shape of the airship is not maintained by gas pressure. In all three types, the cargo and crew cabin is located either underneath or above the aerostatic structure itself.

The Airship as a Freighter. Because static lift is proportional to displacement and displacement (volume) increases faster than structural weight as size increases, cargo capacity rises as a percent of total weight (or volume) as a ship is made larger.[f] The economies of scale of large vessels, which are further augmented by the fact that fuel and crew requirements do not increase as rapidly as size, hold equally well for water-borne or air-borne ships. Hence the same factors that make the supertanker a more efficient ocean-going means of transportation militate in favor of very large airships. The largest zeppelins ever built and flown were 800 feet long with a gas capacity of over 7 million cubic feet, but it may be that present-day versions of these airships will be two or three times as long with up to 15 times as large a gas capacity.[27]

Whereas aerodynamic aircraft tend to be volume-limited in most uses (i.e., the volume available for cargo fills up before payload weight limits are reached), the cargo cabins of huge airships will tend to be weight-limited (i.e., cargo weight limits will be reached before cargo bays are filled). Therefore, bulky, low-density cargo could be carried without energy penalty. Assembled machinery, housing modules, and the like and even airplanes can be transported by airship.

The ability of the airship to hover further enhances its usefulness as a means of freight transportation. Cargos may be carried almost door-to-door, because no elaborate ground facilities are required.[g] Cargos can simply be hoisted up to the ship or lowered from it. This both substantially reduces effective trip time

[f]As a simple illustration of this effect, consider a cube constructed of straight metal bars, each 1 foot long and weighing 1 pound. Twelve bars will be required to construct the cube, thus the structural weight of the 1 cubic foot volume will be 12 pounds. Now if we stretch the cube into a rectangular solid by replacing the four 1-foot bars that connect the ends with four 2-foot bars of the same material, the structural weight will increase by 4 pounds and the new enclosed volume will be 2 cubic feet. Therefore a 100 percent increase in volume has been achieved at the expense of only a 33 percent increase in structural weight.

[g]A mooring post for airships was built near the top of the Empire State Building in the heart of the central business district in Manhattan, but it was never used because of strong and shifting winds.

in normal uses and makes possible the use of airships to transport cargos over rough terrain to remote, undeveloped sites.

The Airship as a Passenger Liner. The relatively low operating speed of dirigibles does not make it feasible for them to compete with large jet aircraft for long-distance passenger travel. However, for short-distance travel airships do represent an attractive relatively energy efficient alternative to the train. Along straight routes airships are not as fast as high-speed trains, but they are faster than conventional trains. Where trains must take circuitous routes because of terrain or the interposition of lakes, etc., airships may have the advantage even over high-speed trains. However, airships also are several times more sensitive to weather conditions.

The roominess of airship cabins may provide another means of attracting passengers away from the faster jet airliners. Passenger airships could be equipped as they once were, with real kitchens and dining rooms, overnight staterooms, lounges, etc. The idea of cruising in relative luxury and comfort on an airship flying only a few thousand feet above the ground and generously equipped with windows might be very appealing indeed. Such a ship traveling from New York to Los Angeles, for example, would provide a breathtaking panorama of the United States while requiring only about one day for the journey.

The safety record of commercial airship operations was respectable. Though more than 60 percent of the zeppelins built were violently destroyed, 63 percent of those were lost in enemy action or to hydrogen fires,[h] 18 percent in ground operations, and most of the remaining 19 percent by unanticipated bad weather encountered in flight.[28] Improvements in meteorological techniques over the past four decades should drastically reduce the latter figure. Nevertheless, the German airship line founded by Zeppelin lost only 13 out of 40,000 passengers carried—all in the famous Hindenburg disaster—and there have been persistent conjectures that that incident might have been the result of sabotage.

There may be some future in short-distance and limited long-distance passenger uses of the airship, but its greatest potential contribution to energy conservation undoubtedly lies in its large-scale development as a means of freight transportation.

Problems in Airship Design and Operation. Since the density of air decreases with altitude, as an airship rises the weight of the air it displaces falls, therefore reducing its static lift. For any given weighting of the craft, there is a maximum height to which it will naturally rise. This is called the ship's *static ceiling.* But atmospheric pressure also decreases with altitude, causing the gas in the

[h]Hydrogen is highly flammable, and would not be used in present day airships.

zeppelin's cells to expand. Therefore, there is also a *pressure ceiling,* the height at which the pressure inside the cells is so great that any further gain in altitude will cause them to burst unless gas is vented. Venting of gas is expensive and causes descent problems and so is used as an emergency measure only.

The zeppelins of the 1930s had pressure ceilings of 5,000 to 6,000 feet and were typically operated at static ceilings of 2,000 to 4,000 feet.[29] Such low-altitude craft cannot usually climb above the weather, implying that the large aerostats have to be designed to ride out the kinds of mild storms they might encounter. They must also therefore be grounded whenever serious storms are occuring anywhere enroute.

Furthermore, as the airship travels it consumes fuel, becomes lighter, and accordingly operates at a higher static ceiling. If the static ceiling is to be maintained, some system for compensating for the fuel weight change must be developed. For the same reason, a hovering airship must take on ballast when unloading cargo and off-load ballast when loading cargo at equivalent rates.

Similarly temperature variations also affect static lift. A cold airship wishing to rise through a layer of warmer air or a warm airship wishing to descend through a layer of colder air might have to wait for the gas temperature to equalize with the air temperature before it could move through the thermal barrier. Related effects from sun/shade variations could also cause problems when the airship is moored.[30]

Although there is no need to have fully developed airports, these airships must have enough clearance to swing in the wind when moored. That may mean that very large areas may be required for maintenance. Likewise even huge ground crews sometimes had trouble controlling the airships of the past. The seemingly most practical current answer to the ground handling problems of these huge behemoths is to keep them continuously airborne from the point of their initial construction and use winches to move supplies and personnel up and down.[31]

Summary and Conclusions

Two of the most important forces encountered when moving an object through a fluid are *drag* (the fluid-dynamic analog to friction) and *lift* or *buoyancy.* The former depends positively on the density and viscosity of the fluid, explaining why the resistance to motion encountered during water travel is so much greater than during air travel and hence why modes of water travel are typically much slower than modes of air travel. The latter may be due to the displacement by the object of fluid weighing more than the object itself (static lift) or due to the pressure effects created by an object moving relative to the wind in accordance with the Bernoulli effect.

Water-borne transportation, both on inland waterways and on the open seas, is highly energy efficient though slow. It is particularly advantageous for the movement of bulky and heavy freight and may be of limited usefulness as a short-distance mode of passenger travel in many of the urban areas in the United States. The effectiveness of inland water transportation could be considerably enhanced by the resurrection and modernization of at least part of the United States canal system, which was built during the first half of the nineteenth century.

Sea-going vessels are presently powered by either low- or medium-speed diesels or by steam or gas turbines. Of the current designs, the low-speed diesels are by far the most energy efficient. Aside from proper choice of engine design, the energy efficiency of sea transportation may be maximized by operating ships at a constant and relatively low speed, by streamlining, by the use of effective anticorrosive and antifouling point systems, and by relatively frequent removal of encrusting organisms from the hull. Recent advances in underwater hull scraping techniques may make the last recommendation more economical by allowing cleaning without drydocking, a time-consuming and expensive procedure. Increasing vessel size, rather than increasing speed, is the most energy efficient means of increasing the rate of cargo delivery.

The attractiveness of inland and sea-going water transportation can be increased by improved materials handling techniques. Among these, one of the most important is containerization. Inland water and open-sea transportation, using whole truck trailers, railroad car bodies, or even barges as containers, and properly coordinated with each other and with modes of land transportation are not only energy efficient but also economical and logistically effective. Hence water transportation can be an attractive and important component of the total freight transportation net.

Jet aircraft are far too energy intensive to be a reasonable way of carrying all but extreme priority, small, light-weight, highly valuable cargo. But they are also far too fast to be replaced as a major means of long-distance passenger travel. Accepting this, we can still maximize the energy efficiency of passenger air travel by eliminating short-distance flights, scheduling fewer flights of much larger jets, avoiding present highly energy inefficient supersonic transport designs, reducing speed of existing jetliners slightly, cruising at optimum altitudes, taxiing with fewer engines or by truck towing, and using improved scheduling techniques and coordination, including ceiling limits on numbers of flights permitted per unit time over congested locations.

In contrast with the energy inefficient jet airplane, the aerostat, particularly the zeppelin, is an attractive mode of freight transportation. Making use of static rather than powered dynamic lift, these airships are not only 5 to 10 times as energy efficient as airplanes but are also large enough to haul great quantities of freight without energy penalty for sheer cargo bulk. Though the optimum cruising speed of previous airships was only about 50 to 120 mph (depending on design), their ability to fly straight without regard to rough terrain, bodies of water, etc. makes them an interesting energy efficient alternative to rail freight over some routes. These supertankers of the air can hover, eliminating the need for elaborate and expensive airports and allowing them to fly routes that are far closer to being door-to-door than those of conventional aircraft.

Airships are restricted to cruising altitudes of only a few thousand feet and hence cannot fly above the weather. They are also subject to buoyancy variations as fuel is consumed, as temperature changes occur, and as cargo is loaded or unloaded. These are all problems, but none appear to be insurmountable.

Overall, water transportation and aerostatic air transportation are (or promise to be) highly effective energy efficient modes of freight transportation. But, because of speed limitations, neither is likely to be of widespread usefulness for long-distance passenger travel. Jet aircraft, on the other hand, are poor (i.e., highly energy inefficient) freighters and should not be generally used for that purpose. But they are so effective as a means of long-distance passenger transportation, that they should be accepted as such and made as energy efficient as possible within that context.

Part III
Industry and the Food System

Industrial Products and Processes

In the early 1970s the industrial sector was the largest energy consumer in the United States, accounting for more than 20 quadrillion BTUs or almost 30 percent of the total.[1] But that energy consumption, rather than being evenly spread throughout industry, was concentrated primarily in six two-digit industry groups:[a] food and kindred products (SIC 20), paper and allied products (SIC 26), chemicals and allied products (SIC 28), petroleum and coal products (SIC 29), stone, clay, and glass products (SIC 32), and primary metals (SIC 33). In fact, the chemicals and primary metals industries alone accounted for more than 40 percent of industrial energy consumption.[2]

Despite the enormous heterogeneity of specific industrial products, only a few classes of energy-use encompass nearly all industrial energy consumption. Two of these, process steam and direct heat, dominate the scene, annually consuming about 70 percent of the industries' and hence more than 20 percent of the nation's energy total.[3] The remainder of industrial energy is used mainly for powering machinery, for electrolytic processes, and for industrial feedstocks (i.e., nonenergy uses of fuels as raw materials, as in the petrochemical industry).

Although we have purposely directed our attention toward the reduction of gross energy consumption and away from the question of usage of particular fuels, there is need for an exception here. As of the late 1960s and early 1970s, something on the order of 45 percent of all industrial energy used was in the form of natural gas.[4] Much of this gas was burned only for the purpose of generating heat and not because of any special property of the gas itself. There are two reasons why this widespread use of natural gas by industry is upsetting. First, natural gas is apparently the shortest in supply (at least domestically) of all the fossil fuels. Second, natural gas is a beautifully clean burning fuel. Its use should therefore be directed with priority (given a short supply) to those situations in which a high degree of flame purity is required or in which uses are so dispersed and small-scale that proper burner maintenance, fuel system optimization, and pollution control are unlikely to be efficiently carried out. However, if natural gas is being used as industrial boiler fuel none of these conditions are met.

[a]According to the Standard Industrial Classification System (SIC) used widely in the United States, industries are classified broadly into a series of two-digit industry groups, which are then further divided into progressively narrower three-digit, four-digit, etc. categories. For example, primary metals is a two-digit industry (SIC 33), while blast furnaces and steel mills is a four-digit industry (SIC 3312) within the broader primary metals group.

The purity of the flame is largely irrelevant to the functioning of the energy system, and the use is so concentrated that proper maintenance and operation, as well as pollution control measures, are both technically available and economically justifiable. On the other hand, residential fuel uses are typically small and dispersed (except in apartment house complexes) taking place in the absence of technical expertise concerning fuel system operation and maintenance. Therefore, if natural gas is to be used as a heat source, the dual objectives of efficient energy-use and pollution abatement would indicate it should be burned chiefly in residences and not in industry.

In the United States energy has not only been cheap per unit, but the energy costs of manufacturing virtually every product have also historically amounted to only a few percent of the total cost of manufacture. As a result of the former fact, whenever it became technically possible to trade off increased energy use for labor (or capital) savings, this has readily been done. As a result of the latter fact, very little attention has been paid in most industries to the flow of energy within the enterprise. Clearly one cannot properly optimize energy-use, just as one cannot optimize the use of money, unless it is first known where it is going, i.e., how and for what purpose it is being used.

This is not to say that no one in the firm knows how much is being spent on energy or that no records of fuel expense are kept. This is clearly not true. But it is to say that frequently the information on energy costs has not been in the hands of those most able, technically and by virtue of authority, to take effective conservation measures. These individuals have often in the past been woefully ignorant of the firm's total energy situation. For example, the chief engineer of a fair-sized chemical plant insisted to visiting energy consultants that fuel usage in the plant was only on the order of 3,000 gallons per month, with a total annual cost of about $12,000 to $14,000. A check of the firm's cost accounts soon revealed that the engineer was correct about the 3,000 gallons, but it was 3,000 gallons *per day* not *per month,* and the annual fuel bill was well over $300,000! Though this particular case may have been extreme, it is probably far more typical of the pre-OPEC oil embargo situation in the U.S. than we might be disposed to believe.[b]

Yet another resultant of the historically cavalier attitude toward U.S. industrial energy consumption has been the low pay and low status attached to jobs relating to industrial energy consumption. Those individuals placed in charge of industrial energy systems, e.g., boiler plants, typically have not been trained to understand the operation of those systems fully and hence often make costly, though well-intentioned, mistakes. For example, consulting engineers visiting the boiler plant of a food factory discovered that the burner was being operated with

[b]The consulting engineer who was the source of this story, an individual with considerable industrial experience, has stated that lack of knowledge of true fuel-use by the operating, engineering, and supervisory personnel in United States industrial plants was the rule among plants visited and not the exception.

65 percent excess air rather than at the 25 percent design point. When the person in charge of the operation was asked about this, he indicated the excess air had been turned up to eliminate smoke in order to comply with pollution control regulations. The consultants put several hundred dollars worth of insulation on tanks storing the preheated boiler fuel and turned down the excess air, producing a 15 to 20 percent saving in fuel consumption, a *several hundred thousand* dollar annual savings in fuel cost, and no smoke.

In the few years since the 1973 OPEC oil embargo, energy has clearly become an issue of greater concern to United States industry. Accordingly, some measures have been taken to overcome these energy information deficiencies. Energy accounting (also called Btu accounting) has become a more common phenomenon, and operating personnel are at least somewhat more alert to the energy situation than was previously the case. But alertness is not a substitute for training and understanding, and unfortunately the resumed flow of OPEC oil has made if far too easy to be once more lulled into believing that serious upgrading of the energy-control skills of operating personnel would require an unwarranted expenditure of time, effort, and money. The plain fact is that the closing of the industrial energy information and training gap that afflicts operating personnel is one of the most important prerequisites to the reduction of energy wastage in United States industry.

Product and Process Design

It has already been argued that the design of an energy-using product, whether it be an automobile or an office building, plays a critical role in determining the energy requirements that will accompany its use. The energy required for the manufacture of this or any other class of product is also dependent upon design—primarily upon the design of the production system. But the design of this system is in turn importantly influenced by the specifications laid down for the product in the course of its own design. Therefore the conservation of energy in industry requires attention to be paid to both the design of the product and the design (and operation) of the production system.

Specifying the Product's Parameters

It is very often possible to reduce the amount of processing, and hence the energy required to manufacture any given product, by taking care not to overspecify the product's parameters. The thickness of coatings, the strength of casings, the design of joints, the machining tolerances specified for components, etc. all can significantly influence the energy cost of manufacture. For instance, designing a joint to permit the fastest possible welding by automated systems can result in at least a 50 percent energy savings over manual welding.[5] Or it may be possible to eliminate the energy intensive welding process entirely for certain joints by revising the specified mechanical properties of the joint.

The specifications related to heat treatment of the product or its components are particularly important to its energy cost. Avoiding overspecification of heat treatments can produce very considerable energy savings. For example, the reduction in specified case hardening depth from 0.062 to 0.052 inch for AISI 8620 steel can reduce the time required for the slow, highly energy intensive heat treating process known as *carburizing* by about 3 hours.[6] Energy may also be conserved in metalworking by designing castings or forgings that require minimum machining and by specifying raw materials sizes as close as possible to those of finished components, thus minimizing both machining and associated scrap production.[7]

The specification of materials is also critical to the energy implications of the product. However, here we must be particularly careful not to engage in too narrow a suboptimization. It is often possible to reduce manufacturing energy requirements for a particular product by altering the choice of materials in such a manner as to produce a more than offsetting increase in energy requirements elsewhere. The switch by the soft drink industry from returnable, reusable glass bottles to nonreturnable bottles and aluminum cans, for example, traded off a reduction in bottle-cleaning energy against a much larger increase in the energy embodied in raw materials. Clearly such changes cannot be considered energy conservative even though they may reduce the energy requirements within the bounds of one particular industry. Similarly, changes that increase the energy embodied in the product of one industry may reduce energy requirements elsewhere by a more than compensating amount. It is quite possible that the increased substitution of high-strength, low-weight plastics and/or aluminum for steel in automobiles may be an example of this sort of change.

Though it requires a somewhat broader view of the energy conservation problem, materials specification is a very important determinate of the total energy implications of any given product. From the viewpoint of energy conservation the basic rule for product material specification should be to stay away from highly energy intensive materials unless the energy savings that would directly result from their use is obviously greater than their embodied energy.

Of course, energy conservation cannot be the primary consideration in choice of materials or for that matter in the specification of any other of the parameters of the product's design. If the product is worth manufacturing at all, it should be designed in such a way as to ensure its safe, reliable, and effective performance. We are patently not arguing for shoddy design but rather for the elimination of energy wasting overdesign. It will, in fact, require much more careful design to hold down requirements to the minimum levels consistent with safety and efficient product performance, but it is an effort with considerable energy conservation potential.

Specifying Production Processes and Equipment

Oversizing of production equipment and overspecification of temperatures, pressures, and processing speeds is probably a far more serious source of industrial energy wastage than even the overdesign of product. It is often the result of a conscientious application of the "principle of brute-force engineering" to the design of production systems. This is the principle by which everything is made hotter, faster, higher pressure, higher capacity, etc. than needed in order to be "on the safe side." Put more succinctly, it is the substitution of horsepower for brainpower. Given the lack of comprehensive investigation into the various facets of this problem, its magnitude and nature are perhaps best illustrated by a few examples:[8]

1. A food processing plant, required to treat its effluent before dumping it into a nearby body of water, was emptying the effluent into a group of half-mile long settling ponds in which bacteria would digest it. The oxygen level in the ponds had to be kept high so that the bacteria would thrive, so the firm installed a series of electric motors with paddles attached to agitate, and thereby aerate the water. This system achieved the desired aeration at the cost of *between 500 to 1,000 installed horsepower.* The same result was subsequently obtained with a replacement system consisting of a network of underwater pipes with air holes in them attached to an air compressor, but the *horsepower requirement was dropped to fifteen.*

2. A plant manufacturing jar lids needed to cure the sealing compound used inside each lid. To accomplish this, the lids were carried along a conveyor belt into a 6-foot tall, 7-foot deep natural gas oven. A heat balance taken on the oven revealed that 45 percent of the heat input was being absorbed by the belt, 48 percent was going out of the stack, 3 percent was going into the walls, etc. and very little of the input energy was actually curing the jar lids. A redesign of the oven, primarily involving reductions in both its excessive total volume and in the distance between the heat source and the product resulted in a *fuel savings on the order of 90 percent.*

3. In order to reclaim a valuable catalyst, this firm was placing the waste mixture of oil, carbon, catalyst, etc. on trays and incinerating it in a natural gas oven hot enough to burn off the impurities and leave a catalyst-rich powder. The trays were filled to a depth of about 6 inches, and the total process typically required *5 hours* to complete. Annual natural gas costs were on the order of *2 million dollars.* By reducing the filling depth of the trays substantially, the waste mixture itself could be directly ignited and burned in a self-sustaining fashion, reducing the processing time to *15 minutes* and *virtually eliminating* the need to burn any natural gas.

It is difficult to judge how typical or atypical these few examples are of the general industrial situation, given lack of detailed and comprehensive analyses of industrial energy systems. One would hope that they represent rather extreme situations, and yet the basic point that they make is clearly a serious one: insufficient attention to the energy implications of production system design can be very costly indeed.

The temperatures and pressures specified for production operations should be held to a minimum. Engines used for powering machinery, furnaces, conveyor systems etc. should all be properly sized and should be operated as near to full capacityas is feasible. Oversized units operated at partial capacity are almost never as energy efficient as units designed for those lower outputs. Nevertheless, the infeasibility of replacing existing overcapacity equipment in the short run (due to economic considerations) should not inhibit the reduction of temperatures, pressures, and speeds to those more realistically attuned to processing needs. In general, energy will still be saved.

Particular attention should be paid to energy intensive processes such as heat treating. The full range of alternative techniques available should be considered and the least energy intensive method consistent with cost and performance requirements chosen. There is a strong interface between the design of product and the design of production system in this selection process, arising from the fact that each of the array of alternative methods may have somewhat different cost and performance implications, as well as different energy needs. For example, the case-hardening carburizing process mentioned in the previous section can be substituted by the far more energy efficient processes of induction hardening, plating, metallizing, flame spraying, or cladding. But induction hardening may be more costly, while none of the remaining processes can match the fatigue strengthening properties of carburizing.[9] Such situations are very common and require the product and production process designers to come together in order to optimize the choice of technique.

Scale Considerations

One of the chief considerations in the design of any production system, whether or not energy consumption is separately evaluated, is that of scale. More specifically, it is important to understand the relationship between the size of a productive unit, e.g., a factory, and the efficiency with which that unit produces. Typically, we are concerned with efficiency as measured (inversely) by the cost per unit of output produced. But there is no reason why we cannot focus on a more purely technological measure of efficiency, such as the energy requirements per unit of output, and ask how this kind of efficiency varies with size as well.

It is commonly observed with respect to the cost measure of efficiency that at least until some limiting size is reached, the efficiency of the productive system tends to increase (i.e., the costs per unit of output tend to decline) as the system is enlarged. This phenomenon is called *economies of scale.* Design within the range of optimum scale economies is a basic rule for maximizing the economic production efficiency of an enterprise.[c]

There is some evidence that economies of scale may exist with respect to energy efficiency as well. Some of these scale economies relate, as usual, to the size of the total productive system or one or more of its subcomponents. The decline in the energy/output ratio experienced by the glass producing industries (SIC 3211, 3221, and 3229) over the period 1947 to 1971 was, for example, at least partially due to the use of increasingly larger furnaces.[10] However, there may also be scale economies related to the size of the lumps of output being processed as units, as is illustrated by the meatpacking industry. In this industry (SIC 2011), a shift toward beef and away from hogs, combined with a growth in the average size of cattle between 1947 and 1967, resulted in a larger average carcass size. Since energy consumption in this industry is more closely related to the number of carcasses processed than to their size, the energy required per pound of meat processed fell.[11]

Where scale economies with respect to energy efficiency exist they should, of course, be exploited. But where larger scale implies a substantial increase in the relative use of energy intensive machinery, smaller may well be more beautiful from the energy conservation viewpoint.

Batch versus Continuous Processing

We define *batch processing* here as the intermittent operation of the main components of the productive system to process units in groups of at least a minimum specified size and *continuous processing* as the continuous operation of the productive system to process a continuous flow of units. The important advantage of the batch method, as far as energy consumption is concerned, is that the components of the production system are essentially always operating at (or near) full load whenever they are operating. On the other hand, the use of the continuous method eliminates the dead time between the processing of successive batches.

Where there is a sufficiently large and steady flow of units, the energy consumed in frequent start-up and shut-down of the processing system between batches will almost always be greater than the energy consumed if the processor is kept running. This is especially true of heat-treating processes. But if the processor is kept running, there is no longer any energy advantage in batch processing. Therefore, as long as the processing demand is reasonably large and stable, capac-

[c]If outputs beyond the range of optimum scale are required they are most efficiently achieved by producing with a number of separate productive units, each of optimum size (as in a multiplant firm). If outputs less than that produced at capacity by a single optimum-sized productive unit are required, some efficiency penalty must be accepted.

ity can be designed at a level such that continuous processing is most energy efficient. Industries in which the introduction of continuous processing techniques have produced gains in energy efficiency include blast furnaces and steel mills (SIC 3312) and bread, cake, and related products (SIC 2051).[d,12]

Batch processing will be more energy efficient for operations in which the flow of units is either intermittent or else is small relative to the minimum energy efficient scale for that class of processor. In such a situation, it will almost always save energy to stockpile the units as they arrive at the processing station, process them all in one batch, and then move them to the next stage. If the scheduling of operations is properly handled, the use of a buffer stock of processed materials will prevent this batch mode from delaying subsequent processing steps. It will however somewhat increase the cost of in-process inventories, though this increase is likely to be relatively small in most cases. At any rate, the energy efficiency of batch processing depends critically on the stockpiling of sufficient preprocessed units to constitute a full capacity processor load. If batch sizes are too small, much of the advantage is lost.

The most straightforward generalization of the decision rule for maximizing energy efficiency in the design and operation of a processing system is thus: If the processing demand can be stabilized at a high enough level to exploit the major energy economies of scale, do so and design the system for continuous operation.[e] If the processing demand can only be stabilized at too low a level to operate at an efficient scale or cannot be stabilized at all, use the batch processing technique. Two important notes: first, high-level stable operation is always to be preferred if feasible; and second, this rule will nearly always maximize the energy efficiency of the system. but there may be other considerations that will constrain its applicability.[f]

Design for Recycling

The problem of the reclamation and recycling of the materials contained within a product that has outlived its usefulness can be greatly simplified if the recyclability of the product is given serious consideration at the point of its initial

[d]Continuous processes for casting, rolling, and finishing steel were introduced in SIC 3312 and continuous baking was introduced in SIC 2051.

[e]Operations can often be stabilized, if they are not naturally stable, by using inventory buffers to allow steady operation at the average rate or by finding uses of the processing system that have demand highs and lows opposite to its original uses.

[f]For example, the periodic shutting down of whole processing lines for extended periods may be called for by this rule where the energy economies of scale are large relative to a stabilized flow of demand. But this sort of intermittent operation may involve excessive labor costs (e.g., from frequent layoff and recall expense or the need to pay higher wages because of job insecurity), and hence more economically efficient, less energy efficient steady, low-level operation may be the only realistic alternative.

design and fabrication. Much of the present difficulty in recycling results from the difficulty in separating out component materials that are closely bound together. If the materials are relatively pure themselves but are interconnected with other different materials, the separation problem can usually be solved by a system of shredding and mechanical classification, though at nonnegligible energy cost. But, if the materials are contained within alloys or are fused together, as in tin coatings on steel cans or copper used as a base coating for the nickel-chrome plating on steel automobile bumpers, the problems of separation are vastly compounded.

To be sure, a product must be designed to efficiently perform the functions for which it is intended. Yet there may be many points at which sufficiently good performance can be achieved by the use of alternate materials, different kinds of coatings, different juxtapositions of components. And, if recyclability is considered a criterion of some importance, the selection from among these alternatives may often be possible in ways that will facilitate materials reclamation while imposing little or no performance penalty.

Packaging

The legitimate functions of packaging are to safely and effectively store and protect the product during its lifetime and to make it readily accessible for use. But the United States has experienced a packaging explosion over the past few decades in which the specifications for the amount and kind of packaging to be used have increasingly lost contact with these basic functions. Excessive packaging has become a way of life and a more and more prominent symbol of our profligate use of our material and energy resources. We have clear plastic wrapped heat-sealed packages of cheese, containing 8 to 16 one-ounce slices, each of which is also individually wrapped in heat-sealed clear plastic. We have large metal aerosol cans containing as much as 90 percent propellant and as little as 10 percent product. We have 4-inch high and 1½-inch diameter self-contained plastic bottles of roll-on underarm deodorant packaged inside totally superfluous cardboard boxes 6 inches high and 3 inches wide— boxes that themselves weigh about 30 percent as much as the deodorant. The list is nearly endless.

Aside from the obvious waste of material resources, this orgy of packaging results in needless consumption of energy for the manufacture and transportation of excess packaging materials. These materials add unnecessary weight and/or volume, thereby increasing the energy consumed in transporting the finished product. Thier contribution to the solid-waste problem typically results in further energy costs during disposal. Furthermore, to the extent that the materials used in packaging the product are either inordinately durable (e.g., some types of plastics) or contain environmentally damaging chemicals (e.g., certain dyes), they exacerbate the pollution problem far beyond their basic contribution to solid waste per se.[g]

Excess packaging may even increase the energy required for direct processing of the product. For example, in the production of frozen prepared dinners and similar specialty frozen products, the nature of the product requires that it be packaged before it is frozen. The packaging then acts as insulation, interfering with the freezing process and increasing energy consumption.[h,13]

The packaging specified for any product should be just sufficient to adequately perform the storage, protection, and safety functions. Packaging for all spoilable goods should be readily resealable to avoid product waste without requiring the use of still more packaging and wrapping by consumers once the original package has been opened. Furthermore, all packaging should be designed to facilitate either reuse (preferably) or recycling. Any special dispensers required should be designed for simplicity, reuse, and ecological compatibility.[i] If packaging were designed within the context of this simple set of conditions, considerable pollution and pure energy waste would be eliminated.

As an illustration of the magnitude of the energy-savings achievable by reducing the amount of packaging to a more reasonable level, the Environmental Protection Agency estimated that if 1971 levels of packaging per capita had been equal to those of 1958, nearly 600 trillion Btu's would have been saved. That amounts to about 12 million gallons of crude oil equivalent per day.[14] And 1958 packaging practices were not necessarily optimal.

Reducing Waste in Industrial Operations

Efficient Use of Process Steam

As of the late 1960s industrial process steam was the third largest end-use of energy in the United States (after transportation and space heating), accounting

[g]There is, for example, a currently ongoing controversy concerning the possible damage to the earth's ozone layer that may be caused by fluorocarbon aerosol propellants. Since the ozone layer protects us from dangerous ultraviolet solar radiation, this is an extremely serious matter indeed. (It always seems odd that the burden of indisputable public proof in these matters is generally on those who take the position that the substance in question is highly dangerous—as if the convenience of aerosol furniture polish was worth the chance of a significantly increased incidence of skin cancer).

[h]In addition, because the product is designed to be heated as a unit, it must have an inner wrapping of material that will transmit, but not be destroyed by, heat. Typically the choice is a foil made of highly energy intensive aluminum.

[i]For example, the finger-pressure pump spray design is equally as effective at producing an aerosol mist as is the pressurized propellant spray can and has none of the potentially damaging ecological features. It also allows maximum volume and weight of the product (as opposed to propellant) within the container and does not run out of gas, leaving inaccessible product locked inside an imposing metal can that cannot safely be opened.

for nearly 17 percent of national and more than 40 percent of industrial energy consumption.[15] Yet much of the energy represented by process steam is wasted for reasons no more complex or sophisticated than leaks, lack of insulation, and failure to fully use the steam and recycle its heat.

Leaks. Leaks of steam are almost pure waste of energy. Even small leaks can add up to considerable excess fuel consumption. For example, assuming a 75 percent efficient boiler operating at 300 psi over a 48-hour work week, the equivalent of 894 gallons of #6 fuel oil will be wasted annually for each 1/16-inch diameter steam leak. If the leak diameter is increased to ¼ inch, the annual fuel loss will rise to about 13,200 gallons, from a single leak![16] Such leaks do not have to result from punching holes in pipes: a simple matter of packing a 3/4-inch diameter valve spindle badly enough to leave only a 0.01-inch space between the spindle and the packing will create a leakage ring whose area is nearly the equivalent of a 3/32-inch diameter hole (a hole half again as wide as that which would leak nearly 900 gallons of fuel oil equivalent per year).[17]

As one example of the energy-saving potential of fixing leaks, General Electric reports that a search-and-seal operation at one of its plant complexes in the Midwest located more than 650 steam, air, and water leaks.[18] Furthermore, General Electric's energy coordinator estimates that in 10 years of his experience 20 percent of the energy-saving opportunities in a typical plant come from eliminating leaks.[19]

Insulation. The value of insulation in preventing heat loss has already been discussed to some extent in our previous analysis of building heating, ventilating, and air conditioning systems. There, two basic rules of thumb concerning insulation were stated: (1) It is better to insulate all surfaces of pipes, tanks, etc. carrying or holding heated (cooled) substances a little than to insulate some heavily and leave others bare; (2) The greater the difference between internal temperature and (desired) external ambient temperature, the more insulation should be applied (i.e., surfaces enclosing hot/cold substances should be more heavily insulated than surfaces enclosing warm/cool substances).

The amount of dissipative heat loss in improperly insulated process steam systems can be (and is) enormous. Although there are no aggregate data on the empirical extent of this loss readily available, some feeling for the problem can be obtained by considering two illustrative examples:[20]

1. A 6-inch diameter steam pipe carrying saturated 350°F steam at 120 psi would lose about 875 Btu/square foot/hour if completely bare. With 2 inches of insulation, the energy loss would be reduced to 75 Btu/square foot/hour. Assuming 100 feet of piping, flanged every 9 feet on the average (because of valves, bends, etc.), the total hourly energy savings resulting from the use of the insulation would be 150,000 Btu. Over a year of 48-hour weeks, the savings would be roughly 375,000,000 Btu. Assuming a 65 percent efficient boiler (and 137,000

Btu gallon of fuel oil), about 4,200 gallons of fuel oil equivalent would be purely wasted each year if there were no insulation on this small amount of piping.

2. A cylindrical storage tank, 20 feet in diameter and 15 feet high, holding hot water at 200°F would lose heat at the rate of 371,000 Btu/hour if bare (assuming the tank sits on the ground and loses no heat through its bottom). If the tank were covered with only 2 inches of insulation, the rate of heat loss would be reduced to about 31,000 Btu/hour, a reduction of more than 90 percent in the rate of energy loss. Assuming a 65 percent efficient boiler, over a year of 48-hour weeks the failure to insulate the tank would result in an energy loss of more than 9,500 gallons of fuel oil equivalent above the loss that would occur if the tank were insulated. This easily preventable pure energy waste would exceed 33,000 gallons of fuel oil equivalent annually if the tank were continuously full (i.e., 168 hours/week).

It is difficult to estimate how severe the underinsulation problem is in industry, but it is apparently a major source of industrial energy waste. One tire manufacturer candidly admits, "There are many *uninsulated* hot lines in the curing systems of older plants. . . ."[21] (emphasis added). And given the historically cheap cost of energy and the comparable problem in commercial and residential buildings, it is difficult to believe that situation is highly atypical.

Extracting Maximum Work and Energy from Steam. During the 1920s, a number of large paper manufacturers demonstrated the technical and commercial feasibility of using industrial process steam to generate electricity. So-called cogeneration of electricity more fully utilizes the energy burned for steam/electricity generation and therefore makes excellent energy conservation sense. However, the intervention of the Justice Department terminated this challenge of the paper manufacturers to the electric power industry.[22]

Whatever the wisdom of the legal decisions of the twenties, it is clearly foolish to continue to fail to fully exploit the energy conservation potential of cogeneration. A 1975 study of cogeneration reportedly estimates that some 43 percent of the industrial steam load is generated under conditions conducive to efficient electricity generation and that full use of cogeneration could save the equivalent of 680,000 barrels a day by 1985 and reduce the capital requirements for electric generating facilities by some 2 to 5 billion dollars per year.[23] Of course, cogeneration can also be achieved by the distribution to industry of the steam produced in the generation of electric power by the central power stations of electric utilities.

The energy contained in industrial process steam can be further extracted in a number of ways. For one, the degraded (i.e., low temperature) steam or condensate can be used for space heating. Although this may extract only a small percentage of the original fuel energy content, it can save considerable energy by reducing the need to burn extra fuel in inefficient space heating systems. The use of this steam or condensate to preheat fuels or other substances to be subsequently heated is a second way of more fully utilizing the contained energy. A third way is to recycle the condensate for use in the boiler. Since the condensate is not only heated but also typically nearly free of contaminating minerals, it makes excellent boiler feed.

Multiple use of industrial process steam makes so much energy conservation sense that is should be made standard practice as rapidly as possible. Its desirability is further enhanced by the fact that use and reuse of industrial steam makes ecological sense as well, aiding in the abatement of air and thermal pollution.

Full Utilization of By-Products

Since by-products are by definition naturally, and often unavoidably, produced along with the main product or products, energy has already been expended to produce them. Therefore, if uses can be found for by-products, along with the resultant reduction in material waste will come an energy saving derived from the use of these low incremental energy cost by-products in place of higher energy cost original products.

In some cases, the by-products will have significant direct energy value and thus fuel potential. For example, coke oven gas, coal tar, coke breeze, and blast furnace gas are produced as by-products of the steel industry's operations and should be fully utilized as fuels there.[24] It has been estimated that about 40 percent of the energy requirements for the manufacture of pulp and paper can be provided by its own processing wastes.[25] One firm reports success in firing boilers with fumes from the paint and varnish kettles used at one of its coating plants, thereby reducing air pollution as well as saving energy.[26]

Where by-products have value both as fuel and nonfuel products, their retrievable Btu content should be balanced against the energy required to manufacture (and perhaps transport?) the nonfuel materials they would replace. Any differential ecological problems (e.g., the release of difficult to filter polluting gases when the material is burned) should also be considered.

The direction of significant industrial research and development effort to finding uses for by-product materials is highly desirable from the point of view of both energy and ecology. History is filled with examples of products, once discarded that were later found to be of real value. In a sense, this is one of the highest uses of human technical ingenuity in industrial affairs.

Adaptive Control of Equipment and Processes. An adaptive control system adjusts the operation or series of operations it controls on the basis of data that it receives by monitoring them. The capability for this kind of fine tuning of industrial processes has been greatly enhanced by the advancement of computer technology and has real promise as a technique for energy conservation. General Electric reports, for example, that using adaptive controls on cutting presses has resulted in a 30 percent energy savings.[27] Advanced computerized energy-use monitoring adaptive control systems have recently come into use in the highly energy intensive primary metals industry.[28]

On-line computer control systems are particularly useful in reducing energy requirements in thermal processes. For example, by using an on-line computer for careful monitoring and adjustment of furnace idling temperatures, temperatures of the charge in the furnaces and between furnaces, and the speed of movement of the charge among other factors in steel reheating operations, one European steel plant was reportedly able to reduce fuel consumption per ton by 25 percent.[29] Similarly on-line computer systems have considerable potential in control of combustion operations. It has been estimated that proper ongoing adjustment of industrial burners, as by an adaptive control system, could save from 5 to 30 percent of the fuel consumed.[30]

Before engaging in an all-out effort to install sophisticated computer control systems, it is wise to bear in mind that much simpler procedures, such as leak plugging, increased insulation, and heat recuperation, are not only likely to result in greater gains in energy conservation but are basic prerequisites to maximizing the efficiency of adaptive control systems. Therefore it is only logical to attend to those matters with somewhat higher priority.

Recycling

Recycling is the natural order of things. The earth has existed with essentially the same complement of elemental material resources for billions of years. Natural processes, in general, and life processes, in particular, continually recycle these resources. All living things are, in a sense, only temporary users of material substances—even those of which they are made. But for the most part living organisms do not in themselves recycle the materials they use. Rather, external natural processes, including a series of other organisms, eventually transform the substances full circle.

At primitive stages of development, human beings also fit neatly into this same pattern. But eventually human ingenuity resulted in the creation of technologies by which material resources could be and were transformed so rapidly into other substances as to outstrip the capacity of natural processes to recycle them. Furthermore, unusual transformations increasingly combined natural materials into substances that could not be readily recycled by natural processes, even apart from considerations of the volume of their production. So materials began to be used up in a more permanent sense than was hitherto possible, and the balance of nature became progressively more hostile to the continued provision of the natural resources and environmental conditions most compatible with the needs of living organisms.

This, combined with the unaltered basic fact of a limited stock of natural materials, has created difficulties that have cumulatively reached proportions that can no longer be safely ignored. Therefore, since human technology created the conditions by which natural processes were rendered incapable of performing the required recycling, it is only logical that human processes and technology be used to either facilitate natural recycling or perform the excess recycling required directly, or both.

Reuse and Reprocessing

Recycling may be defined broadly to include both reuse and reprocessing. *Reuse* means that the object is used beyond its first use either for the purpose for which it was originally designed or for a different purpose that still maintains the integrity of the product. A minimum of treatment, such as washing, is typically required. *Reprocessing* means that the original object is reduced to its component parts and used as a raw material, implying that a great deal more treatment is required. For example, returning glass beverage containers for cleaning and refilling is reuse, while grinding them up, remelting, and refabricating them into one or another glass product constitutes reprocessing.

From an energy conservation viewpoint, reuse is nearly always preferable to reprocessing. There may be cases in which the total manufacturing energy required to produce the product from scrap is less than that required to treat and reuse the product, but they are the exceptions and not the rule. It is more likely, though still quite uncommon, that a sturdier and thus more readily reusable version of a product may require greater total energy over its life cycle than a more flimsy throw-away version. This is, for instance, possibly true in the case of cloth versus paper napkins. In the vast majority of cases, however, reuse is certainly more energy conservative.

It is nearly impossible to estimate the total energy saving potential of reuse, because reuse includes everything from returning bottles for refilling to making children's swings out of old tires and quilts out of old clothes. Since not all reuse involves reentry into the manufacturing system, and since reuse possibilities are so diverse, it is difficult to predict which product demands will be displaced and to what extent. Thus trying to estimate the amount of energy saved by full-scale emphasis on reuse is, as someonw once said, a bit like trying to estimate the number of cockroaches on Manhattan Island—you know the number is large, but you have no idea how large.

One instance of reuse that has been studied in some detail is that of returnable bottles. Bruce Hannon, for example, estimated the total energy expenditure (including materials acquisition, manufacture, transportation, waste collection, etc.) for containing one gallon of soft drink in 16 ounce bottles.[31] According to Hannon nonreturnable glass bottles require about three times as much energy as returnables. Extending the comparison to 12 ounce bimetal (steel and aluminum)

Table 9-1
Relative Energy Savings from Reprocessing as Compared to Primary Production

Product	*Approximate Average Energy Savings per Unit of Production (%)*[a]
Aluminum	95
Copper	90
Steel	85
Titanium	75
Plastics[b]	75
Paper[c]	65
Glass	5

[a]These percentages must be regarded as rough approximations of average savings because, though some are undoubtedly more accurate than others, the assumptions made regarding the quality (and source) of primary ores and scrap, and the choice of recycling procedure will have significant effects on the resultant energy savings achievable. These figures are intended to reflect total energy comparisons and are not restricted to manufacturing energy alone.

[b]The technology for separating and concentrating plastics from mixed waste is not well established. It is assumed here that plastics are separated at the household level and collected separately, especially for recycling.

[c]It is unclear, from the source, to what extent nonmanufacturing energy is included here.

Sources: William E. Franklin, et al., "Potential Energy Conservation from Recycling Metals in Urban Solid Wastes," in: *The Energy Conservation Papers*, Robert H. Williams, ed. (Cambridge: Ballinger Publishing Company, 1975), p. 172; calculations based data given in Citizens Advisory Committee on Environmental Quality, *Energy in Solid Waste* (Washington: U.S. Government Printing Office, 1974), pp. 4–9; Committee on Interior and Insular Affairs, *Conservation of Energy* (Washington: United States Senate, 1972), p. 45.

cans, he found roughly the same energy advantage accruing to glass returnables. All aluminum cans, however, require about four times as much energy as the returnable bottles. As a total, the Citizens' Advisory Committee on Environmental Quality estimated that 225 trillion Btu's of energy could have been saved in 1971 by using returnable bottles instead of throwaway cans and bottles.[32] This is the equivalent of almost 40 million barrels of crude oil or about 1.8 billion gallons of gasoline per year!

The total energy conservation potential of reprocessing is considerably easier to estimate than that for reuse, and it is impressive. The approximate reprocessing energy-saving potential per unit of output for a number of important products and product classes is given in Table 9-1. For primary metals and plastics, highly energy intensive products, the savings are the greatest, ranging from 75 to 95 percent, though the savings for paper are nearly as large. In some cases, depending on the recovery system for obtaining reclaimed glass, it may actually require a few percent greater energy for reprocessing than for primary production, but on the whole reprocessing should produce a small net energy gain. However, this is

Table 9-2
Potential National Energy Savings from Reprocessing Municipal Solid Wastes, 1971[a]

Product (% Reprocessed)	Total Municipal Waste Reprocessed (Million Tons)	Total Annual Energy Savings		
		Btu's (trillions)	Gasoline Equivalent[b] (Million Gals.)	Crude Oil Equivalent[c] (Million Bbls.)
Paper (65%)	25.4	304.8	2,438	53.3
Ferrous Metals (90%)	9.5	114.5	916	20.0
Aluminum (70%)	0.5	112.0	896	19.6
Plastics (25%)	1.1	50.0	400	8.8
Glass (70%)	8.5	21.2	170	3.7
Total	45.0	602.5	4,820	105.4

[a]These savings do not represent potential in the sense of a full reprocessing but rather in the sense of reasonable reprocessing percentages.

[b]Assumes 125,000 Btu/gallon.

[c]Assumes 5,712,000 Btu/barrel.

Sources: Citizens' Advisory Committee on Environmental Quality, *Energy in Solid Waste* (Washington: U.S. Government Printing Office, 1974), pp. 7 and 9; and calculations based on the data supplied in the above source.

much smaller than the energy-savings obtainable by use of returnable glass containers.

As with Hannon's estimates, the figures in Table 9-1 are intended to reflect total energy savings. Therefore, they include transportation, collection, and materials acquisition energy expenditures in addition to manufacturing energy. However, the energy expended in the industrial processing of these products clearly dominates the scene.

The reason why reprocessing offers such dramatic energy savings over the manufacture of virgin product is straight-forward. In the primary metals industries, recycled scrap represents an extremely high-grade ore as compared with existing mineral ores. Similarly, in the plastics and paper industries, the recycled scrap is an input that has already undergone most of the necessary processing steps.

Estimates of the aggregate national energy savings (primarily in industry) that would result from a serious attempt at reprocessing recyclable municipal solid wastes are presented in Table 9-2. The total achievable energy savings shown are equivalent to more than a quarter of a million barrels of crude oil or more than 12 million gallons of gasoline per day! In addition to these enormous energy gains, this large-scale reprocessing would reduce the amount of municipal solid waste requiring disposal by about 35 percent, greatly mitigating the increasingly serious solid-waste management problem facing the nation's cities. There would also be an accompanying improvement in general environmental quality, since, according to the U.S. Environmental Protection Agency, in most cases reprocessing results in considerably less generation of industrial air, water, and solid-waste pollution than does equivalent production from virgin materials.[33]

Urban Solid Waste As Low-Grade Fuel

Nearly 80 percent of the municipal solid waste generated annually in the United States is combustible. The Environmental Protection Agency estimates that the approximate average heat value of these combustibles is about 5,500 Btu/pound[34] on the order of one-third that of #2 fuel oil.[j] Most of this refuse has low sulfur content and can be burned as fuel, under proper conditions, without significant odor or smoke. According to a 1971 study, the combustion of all burnable municipal solid waste in the United States as a supplementary power plant fuel would have reduced power plant coal consumption at that time by 25 percent, with clearly beneficial environmental effects.[35]

One specific product with particularly high heat value is the automobile tire. Each tire, if burned properly yields about 260,000 Btu. Since some 200 million tires are discarded annually in the U.S., the net fuel value of this single combustible item is equivalent to about 1,140,000 gallons of gasoline per day![36]

The potential theoretical fuel-energy value of the total annual flow of combustible trash in the nation's cities is, of course, much larger than this. In 1971 it represented about one quadrillion Btu's—the equivalent of 22 million gallons of gasoline, 20 million gallons of #2 fuel oil, or nearly a half million barrels of crude oil per day.[37] However, from an energy conservation point of view, combustion is inferior to either reuse or reprocessing as a method of managing solid wastes.

In the case of newspapers and other separable paper components of municipal trash, reprocessing saves about 10 percent more energy than burning; for separable plastics, the energy advantage of reprocessing is much greater—on the order of 40 percent.[38] Of course, there are combustible components of solid waste that cannot be readily reprocessed (such as nonseparable plastics and paper, food wastes, wood, etc.), and it is obviously better to burn these and at least reclaim their heat value than to discard them in so-called sanitary landfills or open dumps. However, where there is a choice, reprocessing is always to be preferred. Aside from its considerable incremental energy value, it conserves material resources and, on the whole, reduces environmental pollution.

---

[j]Plastics, the combustible trash category with the highest heat value, averages about 11,000 Btu pound.

There are two main ways in which fuel value may be derived from those combustible products which, having fulfilled their initial purpose, can neither be reused nor reprocessed: direct incineration under conditions in which their heat value may be efficiently used, or treatment by a process designed to transform them into fuels of a more traditional nature. The advantages of the latter procedure lie in the elimination of some of the disadvantages of raw refuse as a fuel.

Raw refuse is typically nonhomogeneous with respect to material composition and heat value. In addition, its moisture content is considerable and varies on a daily basis with the weather (among other things). For example, an analysis of 126 raw-refuse samples taken in St. Louis revealed heat values ranging from 3,027 to 5,809 Btu/pound, and moisture contents of between 9.3 and 50.1 percent.[k,39] However, raw refuse can be shredded and processed in such a manner as to increase its uniformity sufficiently to make its direct use as a fuel practical.

Once the refuse has been properly prepared, there are a number of ways its heat value may be reclaimed.[40] The simplest is to burn it in an incinerator whose combustion zone and lower stack are lined with water carrying tubes that absorb the heat, thus generating steam. Alternatively, the refuse can be burned with coal or oil in the boilers of power generating plants. Another more recent approach is to burn the refuse at high pressure and exhaust the resultant hot gases through a gas turbine-generator to produce electricity.

One of the most interesting refuse transformation processes is *pyrolysis,* in which refuse is heated in an enclosed chamber containing an oxygen-free atmosphere. The organic components of the waste decompose under these conditions yielding char and an oil and/or gas with fuel value. For example, one waste-treatment process which utilizes pyrolysis, known as the *Garrett process,* produces about 160 pounds of char, one barrel of oil, and varying amounts of a low heat value gas per ton of municipal refuse treated (along with roughly 130 pounds of glass, 135 pounds of ferrous metals, and 30 pounds of nonferrous metals).[41] The char has potential value as activated carbon in addition to its heat value, and the gas is usable as a fuel for some purposes (though its energy content is only 400 to 500 Btu/scf), but the oil is the main fuel product of the process. Though it has less than 60 percent the energy value of #6 fuel oil per pound (10,500 Btu/pound versus 18,200 Btu/pound), it is considerably more dense and thus has more than 75 percent the energy value of #6 fuel oil per gallon (113,900 Btu/gallon versus 148,850 Btu/gallon). It also has a sulfur content more than 50 percent lower than low sulfur versions of #6 fuel oil (less than 0.3 percent as opposed to 0.7 to 3.5 percent). Therefore this pyrolytic oil made from garbage is a fairly attractive fuel.[l]

---

[k]This sampling was performed during unusually wet weather and hence likely has an above-average moisture content.

[l]Another pyrolytic waste treatment system that may prove even more interesting is the so-called Purox process. It accepts waste without any of the preliminary shredding and

In addition to pyrolysis, the organic component of wastes can also be converted to fuel by such processes as chemical reduction and anaerobic digestion (biological).[42] One version of the former, developed by the Bureau of Mines, produces a synthetic oil with 85 percent of the heat value of #6 fuel oil–higher than that of the Garrett process. The latter yields methane gas with a heat value of 1,000 Btu/scf and a high-protein residue suitable for animal feed, along with other products. However, both of these processes are designed for treatment of wastes with a much higher organic component than that required by the Garrett system and thus are more suited to agricultural and mixed agricultural-urban wastes.

Industrial Energy Conservation and the Old Plant Problem

Any extensive survey of industrial energy consumption in the U.S. on a plant-by-plant basis will surely show that, in most cases, older plants are less energy efficient. The reasons for this are two-fold: greater physical depreciation combined with inadequate maintenance and embodied technological progress.

All buildings, machines, and equipment wear out with age and use. Accordingly, their need for maintenance increases with the passage of time. Maintenance, however, is typically a labor intensive process, and given the high labor costs in the United States, a fairly expensive one. In addition, the effects of less than optimal maintenance are generally not immediate, so there is always a temptation to follow a short-sighted policy of undermaintenace in order to hold down costs in the short run. Therefore, for reasons of expense and myopia, maintenance is nearly always below technically optimum levels and often below long-run economically optimum levels as well.

As a result, older plants tend to have a larger fraction of equipment performing below design levels. Beyond this, the historical cheapness of energy in the United States has meant that malfunctions whose major result was to waste energy have been given least priority, even if they called for relatively simple maintenance. The result is that older plants have more steam leaks, broken windows, worn out insulation, badly adjusted boilers, etc. than newer plants and thus are far less energy efficient.

Technological progress is often classified as either embodied or disembodied. The embodied form is incorporated into the design of a new piece of equipment; the disembodied form is not. For example, the development of numerically con-

---

drying required in the Garrett process or any other preparation, for that matter. The pyrolysis stage takes place at much higher temperatures and produces a clean gas as its main fuel product. In addition, there are reportedly virtually no stack emissions. (See Dov Iliashevitch, "Municipal Solid Wastes as a Source of Energy Conservation," pp. 39–42.)

trolled machine tools was an embodied technological advance, while the development of the industrial engineering technique known as PERT for scheduling and control of projects (such as building construction) was disembodied. The importance of this distinction is that disembodied technological progress may be adopted and used with existing equipment, while the embodied variety requires the purchase of the new equipment incorporating the development. Hence, to the extent that more efficient technologies are of the embodied type, older plants, almost be definition, will be restricted from utilizing them.[m] They will, however, be built into new facilities.

In an impressive series of industries, embodied technological advances that have improved the general efficiency of operations over time have also improved their energy efficiency. The so-called Pilkington float glass process for manufacturing plate glass (perfected in the late 1950s) for example, uses all productive inputs more efficiently, including fuel.[43] Likewise, the basic oxygen process (perfected in the mid-1950s) has low installation costs, produces high quality steel, and is also more energy efficient than the open hearth process it is replacing.[44] Cement manufacture is yet another industrial activity in which advances in process, mechanization, and control have reduced energy requirements per unit of output while improving overall productive efficiency.[45] Therefore, despite the cheapness of energy and the inattention to energy conservation, embodied technological advances adopted for other purposes have often rendered new plants more energy efficient.

Historically, in industries in which energy was an important cost factor direct pressures for the improvement of energy efficiency have resulted in embodied technological advances that have widened the energy efficiency gap between old and new facilities. One of the more interesting examples of such energy efficiency directed technology is the new Alcoa smelting process, which promises to reduce energy consumption in the highly energy intensive manufacture of primary aluminum by 30 percent.[46]

If older plants are usually less energy efficient, does this mean that we must wait 15, 20, 30, or more years for the natural replacement of older facilities before we see a real mollification of the old plant problem? Or does it mean that we should engage in a huge-scale wrecking and replacement program for industrial facilities to speed up the process? What is to be done?

In the first place, we must remember that much of the old plant problem is due to the underutilization of rather standard renovation and maintenance procedures, like replacing insulation (or insulating previously uninsulated

---

[m]Of course, the term older plants does not refer simply to the factory buildings, as such, but much more importantly to the complex of equipment and facilities contained therein. It is often, though not always, possible to incorporate new processes or install newly developed equipment in old buildings. In such cases the physical building deterioration will still tend to make the factory less energy efficient, despite the fact that the new equipment does embody technological advances and accordingly mollifies the energy inefficiency problem. Typically, however, old plants and old equipment are correlated.

pipes), fixing steam leaks, adjusting furnaces, etc. Accordingly, there is no reason why considerably improved maintenance cannot greatly increase the energy efficiency of older plants at a relatively modest cost. In the second place, the premature replacement of facilities in itself requires excess energy expenditure for the scrapping of old plant and equipment and the manufacture and construction of new facilities. Wholesale replacement of older plants could therefore produce a net increase in energy consumption. But even more importantly, these new plants would eventually get old, and unless one was willing to embark on the extraordinarily expensive, resource-wasting and energy inefficient (and hence self-defeating) policy of replacing all industrial facilities every few years, the old plant problem would merely be postponed, not solved.

The optimal procedure, from an energy conservation point of view is to follow a policy of encouraging increased energy relevant maintenance (particularly of older facilities), on the one hand, and the more rapid replacement of technologically obsolescent equipment with more energy efficient designs, on the other. It might be wise to supplement the natural economic pressures deriving from higher fuel prices with additional artificial incentives, such as tax breaks (tied specifically to energy efficiency increasing activities), in order to encourage improved maintenance and the installation of energy-saving systems more or less compatible with existing facilities (e.g., on-line computer furnace controls). If, as Berg reports, loans for expanding plant capacity seem to be given higher priority than loans for improving the performance of existing plants, removal of this bias would also be helpful.[47] A system of specialized tax or other governmental incentives for the replacement of existing equipment with new, more energy efficient facilities might be in order. Special incentives for the development of energy efficient innovations might also be incorporated into this approach to the old plant problem. It should be noted, however, that the incentives applied do not necessarily have to be positive (e.g., tax reductions) but could be negative (e.g., tax penalties) or a combination of both. The only requirements are that they be effective and that they do not stimulate the replacement of facilities beyond the point of net energy gain.

Summary and Conclusions

Industry is the largest energy-consuming sector in the United States. Much of the energy consumed by industry is in the form of natural gas used purely as a heat source. This particular energy resource is too short in supply and too clean-burning to be used where dirtier fuels will suffice and hence should be directed to those (primarily residential) uses where it will produce the greatest net pollution reduction and energy efficiency increase.

The historical cheapness of fuels in the U.S. and their relatively minor role in product cost have resulted in an overly casual attitude toward industrial energy consumption. As a result, record keeping on energy-use and the training of personnel responsible for operating large energy-using components of industrial processes both leave much to be desired. The correction of this inordinate energy ignorance is fundamental to the successful conservation of energy in industry.

Product design has important implications for the energy required for manufacture. Care should be taken not to overspecify, particularly with respect to product parameters relevant to heat treatment or other highly energy intensive processes. Similarly overspecification of temperatures, pressures, and processing speeds and oversizing of production equipment should be avoided in the design of production systems. Choice of materials is also critical to the product's energy implications, but care must be taken to consider not only the manufacturing energy required within any given industry but also the energy already embodied in the materials used. Careful product and process design is enormously important to industrial energy efficiency.

There is some evidence that the scaling of industrial facilities and of units of product may have significant energy implications. There is also some evidence that continuous processing is more energy efficient than batch processing wherever there is a smooth flow of product at a large enough level to ensure high-capacity utilization. However, batch processing is more energy efficient for subcomponents of the productive system where the demand for that particular processing step is either intermittent or else small relative to the minimum energy efficient scale for the operation. As a general rule: if processing demand can be stabilized at a high enough level to exploit the major energy economies of scale, it should be done, and the system should be designed for continuous operation; if the processing demand can only be stabilized at too low a level to operate at efficient scale, or cannot be stabilized at all, batch processing should be used.

Excessive packaging is responsible for a great deal of energy waste. It absorbs energy unnecessarily in its own manufacture and transportation; it adds to the weight and volume of the product, thereby increasing fuel requirements for product transportation; and in some cases, it may even increase the energy required to manufacture the product itself. The reduction of present packaging levels to those of 1958 would, by itself, save the energy equivalent of between 0.25 and 0.33 million barrels of crude oil per day.

Process steam is a major component of industrial energy consumption. A large amount of the energy contained in process steam systems is wasted because of nothing more sophisticated than unattended leaks and insufficient insulation. The energy embodied in industrial process steam could also be more fully utilized by using it to generate electricity in addition to its direct use. It has been estimated that full use of cogeneration could save the equivalent of

680,000 barrels of crude oil per day by 1985. Energy waste could be further reduced by using degraded steam or condensate for preheating, space heating, or boiler feed.

Uses can often be found for industrial by-products, either as fuels or as products with low incremental energy costs. Full use of by-products is not only energy conservative but also resource conservative and pollution reducing. Research into by-product uses should therefore be encouraged.

The advent of small, reliable computer systems adapted to on-line control of energy intensive operations has important energy conservation potential. There is some evidence that such systems can provide significant energy savings when applied to heat-treating processes, and their application elsewhere holds some promise. But the simpler leak-sealing and insulation procedures are not only likely to produce greater net energy gains but also to be prerequisites to the efficient functioning of these more sophisticated computer systems.

Recycling is a very important means by which very large reductions in net energy consumption (particularly in industry) may be achieved coincident with substantial gains in the reduction of air, water, and especially solid-waste pollution. The highest form of recycling is reuse, as typified in the glass industries by returnable bottles. Because reuse requires no significant transformation and little treatment, it is highly energy conservative. It has been estimated, for example, that using returnable glass bottles instead of throwaway cans and bottles would save some 40 million barrels of crude oil equivalent per year.

After reuse, the next most energy conservative form of recycling is reprocessing. Recycled scrap is essentially a very high grade ore and hence requires far less processing than primary materials. For primary metals and plastics, reprocessing would save between 75 and 95 percent of the energy required per unit of output to manufacture the product from virgin materials. As a total, a serious effort at reprocessing could save more than a quarter million barrels of crude oil per day. It would also reduce the municipal solid-waste disposal burden by more than one-third.

Third on the list of energy conservative reclamation measures is use of combustible wastes as fuel. Full use of combustible municipal trash as fuel would produce the equivalent of about a half millon barrels of crude oil equivalent per day. Combustion may be accomplished by burning the materials directly in a water-wall incinerator to generate steam or by mixing prepared refuse with coal or oil and burning it in the boilers of electric utilities to generate electricity. The trash may also be converted into more conventional fuels, such as synthetic oil and gas, by such processes as pyrolysis, anaerobic digestion, and chemical reduction.

As a general rule, the maximum reduction of product energy consumption and pollution will be achieved if anything that can be directly reused is reused, anything that cannot be reused but can be separated is reprocessed, anything that cannot be separated is burned as a fuel, and anything remaining is buried in a sanitary landfill. Careful attention to so-called postconsumer use of the product during its initial design can facilitate recycling by increasing the sturdiness of the product in order to allow considerable reuse and by simplifying the ultimate separation of components to expedite reprocessing.

Older industrial facilities are typically less energy efficient both because they have undergone greater physical depreciation and because they embody older and often less energy efficient technologies. Greatly improved maintenance procedures particularly at older plants, combined with increased incentives to purchase new energy conservative equipment, is probably at once the most practical approach to this problem and the approach likely to produce the largest, long-term net gains in energy conservation.

In 1973 Harvey Morris, the President of a New York-based energy consulting firm, a man with more than a decade of experience as an industrial energy efficiency expert, estimated that U.S. industry could easily reduce its energy consumption by a minimum of 30 percent with a serious ongoing conservation effort.[48] This would amount to a savings of nearly 10 percent of the nation's total. Nothing in our analysis of the potential for energy conservation in industry would indicate that this is an overly bold estimate.

10 Food

Despite the long-term secular decline in the percentage of United States population living in rural areas, the United States food system is still both a major employer of labor and a major consumer of other productive resources. The food system has however grown much more complex, and the distribution of its employment and resource use has shifted toward its nonfarm components. On the supply side, these may be divided into the farm support (agricultural machinery, fertilizers, pesticides, etc.), food processing (meat packing, food canning, food freezing, etc.), food distribution (wholesaling and retailing) industries and that segment of the transportation sector engaged in carrying food and food products. As a total, it has been estimated that some 20 percent of the nation's workforce is involved in one or another aspect of the supply of food.[1]

But a comprehensive definition of the food system would have to include the food appliance (stoves, refrigerators, dishwashers, etc.) industry and the commercial and residential transformation of food and related products into meals. It is this expanded definition that shapes the pattern of this chapter.

The food system in the United States is a heavy energy user. The impressive success of United States agriculture in increasing yields per acre has been achieved at the expense of a massive increase in energy inputs, some direct and some embodied in petroleum-derived fertilizers, pesticides, and the like. Similarly the food processing industries (SIC 20) are among the six high energy-using major industry groups, accounting for nearly 10 percent of the energy purchased by that set of industries in 1971.[2] The final preparation of meals in residences and commercial institutions is also a major point of energy consumption. By the roughest kind of estimate, the energy consumed by food-related appliances in these sectors is about 5 percent of the nation's total.

Table 10-1 gives a nearly comprehensive breakdown of the energy-use in the United States food system for the year 1970. The food system, as a whole consumed nearly 13 percent of the nation's energy, but less than one-quarter of this consumption was directly agricultural. The remainder was more or less equally split between postfarm processing and transportation and food preparation. In the agricultural segment, direct fuel for farm machinery dominates with nearly 45 percent of the agricultural energy total, followed by fertilizer, the chief form of embodied energy, with almost 18 percent. The energy significance of packaging in industrial food processing is illustrated by the fact that it absorbs some 40 percent of the processing energy total. Transportation also plays a

Table 10-1
Energy Use in the U.S. Food System, 1970

	Energy Consumption	
Component	*Trillion Btu's*	*Percent of U.S. Total*
Agriculture	2088	(3.1%)
Direct Fuel	921	
Electricity	253	
Fertilizer	373	
Agricultural Steel	8	
Farm Machinery	317	
Tractors	77	
Irrigation	139	
Industrial Processing and Transportation	3341	(5.0%)
Food Processing Industry	1222	
Food Processing Machinery	24	
Paper Packaging	151	
Glass Containers	186	
Steel Cans and Aluminum	484	
Fuel for Transport	980	
Manufacture of Trucks	294	
Commercial and Home Preparation	3191	(4.8%)
Commercial Refrigeration and Cooking	1044	
Refrigeration Machinery	242	
Home Refrigeration and Cooking	1905	
Total for U.S. Food System	8620	(12.8%)

Sources: adapted from John S. Steinhart and Carol E. Steinhart, "Energy Use in the U.S. Food System," *Science* (April 19, 1974), p. 309.

major role, consuming nearly 15 percent of the food system total. Finally, home food preparation is close to twice as large an energy consumer as commercial preparation, accounting for more than 20 percent of food system energy use.

Agriculture

The series of alterations in agricultural techniques known as the *green revolution* has produced striking increases in the yield of both animal and plant crops. For example, the yield of corn, one of the most important grain crops in the United States, grew from 34 bushels per acre in 1945 to 81 bushels per acre in 1970, an increase of nearly 140 percent.[3] But the green revolution has depended so heavily upon increased energy inputs that the energy efficiency of agriculture, as measured by the ratio of energy value of output to the energy inputs required, has decreased over the same period. For corn this decline has amounted to nearly 25 percent.[4]

There had, in the past, been high hopes that the export of the green revolution to third world countries would play a major role in solving the world food problem. However, the green revolution has run hard up against the energy crisis, and these hopes are rapidly fading. The energy intensivity of modern U.S. agriculture is simply too great to be practical for most third world nations. For example, in order for the average daily caloric intake of the people of India to be upgraded 50 percent to the average United States level using a U.S.-type food system, more energy would be required than India now consumes for all purposes combined.[5] For a nation without large domestic energy resources, this degree of agricultural energy consumption would put a backbreaking burden on its economy by creating a large foreign-trade deficit. In fact, much of the most severe economic hardship created by the sharp increases in petroleum prices in the past few years has been suffered by third world nations that have at least partially adopted green revolution techniques.

The magnitude of the drain on world energy resources that would follow widespread adoption of modern U.S. agricultural procedures would be very large indeed. If petroleum were the only source of energy for agriculture, the entire known world reserves of petroleum would be completely consumed for agricultural purposes in only 29 years.[a]

What then is to be done? Is the green revolution to remain the private preserve of the United States and other developed nations? Can even they continue to afford the enormous energy expenditures the green revolution requires indefinitely? In this section we consider some of the possibilities for reducing agricultural energy requirements so as to make modern agriculture more energy efficient and thus more affordable, without making it significantly less productive. This seems the only equitable way to reconcile and resolve the worldwide food and energy problems.

Selective Breeding

One of the major thrusts in the development of modern agriculture has been the increased use of selective breeding to genetically redesign plants and animals. There is nothing essentially wrong with this idea, but it does have limitations that should be clearly understood. In the first place, so-called improved varieties of plants and animals are not typically improved in all aspects. Rather, they are merely more specialized. Increased specialization usually does produce enhanced performance in the area of specialty but only at the expense of reduced

[a]Considering all estimated potential petroleum reserves would extend the period for resource exhaustion to 107 years. These estimated depletion times are, on the one hand, overstated because they assume no petroleum use other than for agriculture and, on the other hand, understated because they do not consider any other energy resource. (See David Pimentel et al., "Food Production and the Energy Crisis," p. 448.)

performance elsewhere. Thus plants and animals bred for high yield of products that we consider valuable (e.g. fruits, milk, wool) may well lose some of their natural ability to propagate or to withstand unusual climatic variations or even diseases. They become dependent on human intervention to perform those services they can no longer perform for themselves and to otherwise keep their environmental conditions within narrowed limits of acceptibility. For example, corn can no longer shed its seed to propagate itself, and grapes, roses, and citrus fruits must be routinely grafted onto wild rootstocks to increase their vigor.[6]

Secondly, the application of selective breeding often leads to genetic uniformity. Favored varieties of animals and plants are widely grown, so that if some virulent new disease or pest appears on the scene, the chances for species survival are greatly reduced over what they would be if the species were more genetically diverse. Some examples of disastrous plant epidemics that have occurred in the U.S. are the southern leaf blight of corn, the fungus disease of an oat strain known as Victory, and the rapid spread of a new wheat stem rust among a variety of wheat bred for resistance to an earlier troublesome variety of wheat rust.[7]

Plant crops have for sometime been genetically designed to maximize their ability to produce high yields given large fertilizer inputs. In the absence of this heavy fertilization, these plants will not produce high yields, and thus whatever sacrifices in plant vigor or independence have been accepted in order to achieve high fertilizer utilization efficiency will have been suffered in vain. Accordingly, the use of green revolution plant varieties is nonsensical unless they can be grown under the proper conditions. To the extent that these conditions include heavy fertilization, high levels of energy input may be unavoidable if artificial fertilizers must be used.

The economic advantages conveyed by the mechanization of agriculture have led to the development of special varieties of fruits and vegetables bred to withstand mechanical handling, as well as longer-distance shipping. Unfortunately, plant varieties designed for firmness and toughness of skins are often less tasty than more standard varieties and may be less nutritious as well. Thus some aspects of food quality have been sacrificed in order to improve the compatibility of agricultural products with energy intensive mechanization.

It would seem that the idea of selective breeding could readily be turned to the purposes of energy conservation if the breeding objectives were somewhat altered. For instance, although agricultural crops are typically bathed in free, nonpolluting, renewable solar energy, they are only able to capture a small fraction of that energy. Photosynthesis, the process by which plants use light to energize the synthesis of raw materials into food, is only about 1 percent efficient.[8] Perhaps it is conceivable to develop high-yielding varieties of agricultural plants that are somewhat more efficient at capturing and utilizing this free energy. Plants might also be bred for increased water-utilization efficiency (to reduce highly energy intensive irrigation) or increased insect and disease resistance (to reduce the need for artificial pesticides). Since it is not likely to be possible to achieve all these objectives simultaneously, it is important to review the particular energy characteristics of the agriculture of any given crop in order to identify which of these or other breeding objectives have the greatest potential for energy reduction and hence should be accorded priority. By way of illustration, present varieties of corn have much higher moisture content at harvest than older strains. As a result, drying them for storage required 12 times as much energy per acre in 1970 as it did in 1945 and in fact was the third largest use of energy in corn farming.[9] Developing reduced moisture varieties of corn might therefore produce substantial energy savings.

The search for new varieties of plants with desirable food and energy-relevant characteristics should not be restricted to variants of those species already cultivated on a large scale. There are only 50 species that are presently major agricultural crops, 12 of which supply 90 percent of the world's total production. But there are about 80,000 edible species of plants, and it would be unwise to ignore all of the possibilities these represent.[10]

Natural Fertilizers

Since present crop varieties require extensive fertilization in order to produce the high yields of which they are capable, it is important to consider what alternatives there are to highly energy intensive artificial fertilizers. There are at least two natural fertilizers available that could provide sufficient nutrients to the soil to allow high-yield crop varieties to flourish, while at the same time producing very substantial savings in agricultural energy. One is livestock manure and the other is a class of soil-enriching plants that have been called *green manure.*

United States livestock manure production is on the order of 1.7 billion tons per year, more than half of which is produced in feedlots and confinement rearing situations.[11] Since the runoff from animal feedlots is a serious pollution problem in some parts of the United States, the recycling of feedlot livestock manure as fertilizer would not only reduce the energy costs of agriculture but also ameliorate the ecological difficulties associated with feedlot operations. Once again, it is clear that the conservation response to the energy problem complements, rather than conflicts with, the objectives of pollution reduction.

It has been estimated that the substitution of manure for chemical fertilizer in corn production would save almost 4.4 million Btu per acre.[12] Assuming the availability of only 20 percent of the manure produced in feedlots for corn fertilization, some 17 million acres (30 percent of 1970 corn acreage) could be treated.[13] The resultant energy savings would be more than 74 trillion Btu per year, the annual equivalent of about 13 million barrels of crude oil.[b]

---

[b]Beyond the direct energy and pollution control benefits of using manure as fertilizer,

An alternative to livestock manure fertilization is the planting of legumes or other appropriate crops in rotation with the main crop. This simple procedure is not only highly effective in adding nitrogen to the soil but may also have pest and disease control advantages.[c] For example, planting sweet clover in the fall and plowing it under after 1 year adds considerably more nitrogen to the soil than is required for corn growth while at the same time helping to control the corn rootworm and reducing disease and weed control problems.[14] When full-scale rotation is not feasible, similar advantages may be achievable by planting appropriate alternate crops in the off season. Seeding corn acreage with winter vetch at harvest time and plowing it under in the spring still adds more nitrogen than the corn crop typically requires, though it is not as advantageous as annual rotation.[15]

The energy savings achievable by replacing chemical fertilizer with green manure are even greater than when livestock manure is used. For an acre of corn, the estimated savings are almost 6 million Btu per acre.[16] Assuming equivalent acreage to that in the livestock manure case (about 30 percent of total harvested corn acreage in 1970), the energy savings from the use of green manure in corn production would amount to more than 100 trillion Btu or the equivalent of some 17.7 million barrels of crude oil per year. But there is no reason why the savings cannot be several times as large, since green manure is subject to neither the supply limitations nor the same logistical problems as is feedlot manure.

Here, as with primitive architecture and flywheels, we see an example of a situation in which a return to older, "outmoded" principles promises large energy savings without significant sacrifice in performance. There is no reason why the use of either brown or green manure should force a substantial sacrifice in agricultural yield, and yet their energy savings potential is considerable. A recent comparison of a sample of operating organic farms that used such fertilization practices exclusively reportedly revealed that their yields of corn were about 10 percent lower than conventional farms, but that the yield reduction for oats was slight and the yields of soybeans and wheat were about the same. However, the organic farmers used only one-third as much fossil fuel energy to produce the same crop quantities and made as much money as did comparable conventional farmers.[17] Thus what limited evidence there is suggests that organic farming is both economically viable and highly energy conservative.

Agricultural Machinery and Systems

Machine Scale. As in industry, the scaling of machinery and equipment may be excessive in agriculture as well. Since direct fuel and machinery account for

---

the organic matter it adds to the soil increases the number of beneficial microorganisms, improves the soils resistance to erosion, and increases its water holding and percolating capacity. (See David Pimentel, et al., "Food Production and the Energy Crisis," p. 446.)

[c]This green manure has the same advantages as livestock manure in adding organic matter to the soil.

a large portion of energy consumption on the farm, using smaller tractors or other machinery and somewhat more labor could produce substantial energy savings without reducing yields. Farm machinery, like industrial machinery, is more energy efficient if it is more precisely scaled for its work and operated continually at high capacity.

In third world nations, the advisability of using smaller-scale agricultural equipment is heightened by the combination of high fuel and equipment costs and low labor costs. While extremely labor intensive methods may not be capable of producing large agricultural yields, it is unlikely that highly capital intensive methods will result in tolerable food production costs. It would seem, then, that agricultural technologies somewhere between primitive labor-oriented and modern fuel-guzzling capital-oriented technologies would offer the best compromise for the third world.[18] For the United States, the optimal agricultural technology would likely still be highly capital intensive, but the concerns of energy conservation do call for some degree of backing-off from extreme agricultural mechanization.

Weed Control. It is possible to reduce the energy consumed in the control of undesirable plants by substituting mechanical cultivation for mechanized spraying of herbicides. The savings, however, are likely to be relatively small. In the case of corn, for example, energy-use for this purpose would be reduced by 5 to 10 percent.[19] But, if hand spraying of herbicides is used in place of mechanical spraying, very large relative energy savings should be produced. Estimates for corn production indicate a savings of more than 98 percent.[20] Neither of these substitutions should produce any significant reduction in the efficiency with which weed problems are managed, though the greatly increased labor content of hand spraying might well make it seriously uneconomical for widespread use at current labor and fuel prices.

Irrigation. The dream of vastly increasing the world's cultivable land through mechanical irrigation (and other environmental manipulations) has come into conflict with the global energy problem. Mechanical irrigation is a highly energy intensive process. Pimentel, Dritschilo, Krummel, and Kutzman have estimated, using the water requirements of corn as their basis, that suggestions of doubling the world's arable land by irrigation imply a need for more than 120 quadrillion Btu per year, just for irrigation of this new land.[21] This is 70 percent more energy that was consumed in the United States for all purposes in 1972 and, if supplied solely by petroleum, would deplete the world's known usable petroleum reserves in about two decades!

Fortunately, there may be a way to make the deserts bloom without mechanical irrigation by reaching back once more into past human technology to revive the principles of an ancient technique known as *runoff agriculture.*[22]

During the period from the end of the third century B.C. to the middle of the seventh century A.D., the practice of runoff agriculture reached its peak in the Negev desert in the south of Israel. In the Negev, annual rainfall averages only three or four inches, and yet ancient farmers in this region were able to grow crops whose water requirements were several times this high without any machinery. The basic idea was to use catchments many times larger than the cultivation area with channels in their sloping sides that would direct the runoff from whatever rain fell to the cultivation areas. Any given catchment would always be higher than the growing area it served (usually located in a valley or on a flood plain), so that the runoff was transported to the desired area gravitationally. If the catchment is 30 times as large as the cultivated area and only one out of four inches of rainfall becomes runoff that reaches its desired destination, then the growing area receives the equivalent of an annual 34 inch rainfall—enough for virtually any crop or orchard!

In 1959, experimental farms using these principles were established in the Negev, at least one of which was built with steel pipes, concrete channels, and other similar accoutrements of a modernized version of the floodwater distribution system. Over a 14-year period, during which annual rainfall varied from 1 to 7 inches, considerable success was achieved. Attempts to grow peaches, apricots, pistachios, apples, almonds, cherries, plums, grapes, etc. resulted for the most part in average to good yields. Vegetables, field crops, and pasture plants were also grown successfully.

Furthermore, in 1961 an experiment with a modified system of microcatchments was begun. Each fruit tree, shrub, or other plant had its own small catchment rather than all being serviced by a single large catchment. The experiment was highly successful. The microcatchments collected as much as 62 percent of the rainfall as runoff, allowed runoffs in rainfalls too small to produce any runoff from the larger catchments, and could be installed and maintained at very low costs.

It is certain that soil conditions and other related matters will influence the effectiveness of runoff agriculture. More experience may be required to define its limits. But the main point is that runoff agriculture does work, and it represents a highly energy conservative and ecologically beneficial alternative to mechanical irrigation.[d]

Food Processing

There is simply too much processing of food in the United States—too much from the viewpoint of energy, too much from the viewpoint of nutrition, and too much from the viewpoint of taste. In many cases, we have succeeded in making foods look better, look more uniform, by various physical and chemical treatments at the same time we have reduced their palatability. So we have, for

---

[d]The ecological benefits derive from reducing flash flood damage and ultimately from restoring the land.

example, reconstituted potato chips, unbroken and uniform in shape and color, but with little taste. And we increasingly find evidence that many of the chemical substances we had been blithely adding to our foods for years may be linked with maladies ranging from hyperactivity to cancer. One does not have to be an advocate of health foods to realize that we have gone too far.

Even those foods we think of as being fresh frequently do not deserve the name. For example, in recent years it has become more and more difficult to buy a fresh apple. Apples are nearly always stored for long periods after they have been picked in cooled sealed environments filled with 5 percent carbon dioxide atmospheres to prevent rotting and older apples are normally sold first.[23] The apples typically available at food markets are accordingly beautiful looking but with tough skins, and mediocre consistency and taste.[e] Bananas are another example of processed "fresh" fruit. They are picked green to make them more amenable to long-distance shipping and then ripened artificially in storage houses by being exposed to ethylene, a common hydrocarbon gas.[24] Any individual who has been living in an urban area of the United States (and, to a lesser extent, even in rural areas) long enough to have forgotten what fresh fruit tastes like will be struck by the contrast in taste between what now passes for fresh fruit in much of the U.S. and those fruits still readily available in some other countries.[f] But it was not always thus in the United States, and we need not continue to accept this energy intensive degradation of our food supply.

One of the particularly energy intensive food processing innovations of recent years was the development of the frozen prepared dinner. The life-cycle of this product is almost a study in how to consume energy. First the components of the product are individually prepared, then typically at least some of them are cooked. The entire dinner is next assembled, packaged, and frozen. From this point on additional energy must continually be consumed in order to keep the product frozen during transportation and storage. The dinner is often stored in open freezers in retail outlets—freezers that spill cold air constantly, requiring additional make-up energy to chill new air. After purchase, the dinner is transported (in small inefficient loads) to residences where it must be kept frozen until it is desired, at which point it must be reheated and then consumed, leaving a lot of energy intensive aluminum and other packaging to be thrown away. Each step of this process depends on continuous energy consumption.

And what do we get for all this energy expenditure? A partial list of ingredients for one such product includes (in descending order by amount): reconstituted dehydrated potatoes, cooked turkey meat, margarine, reconstituted whole milk, turkey skins, turkey fat, modified food starch, apple powder, de-

---

[e]In the fall of 1974 the author went apple picking in an orchard about 50 miles or so from New York City. The taste of the apples picked, even those that did not look all that good, was almost unbelievably better than that of the store-bought variety. And this was still true after 6 to 8 weeks of refrigeration.

[f]The peaches available in Switzerland are just one of many, many examples.

hydrated onions, flavoring, coloring, and ascorbic acid (along with some more appetitzing ingredients). Although this product would appear to be a well-balanced meal of meat, bread, potatoes, fruit sauce, and vegetables, the manufacturer recommends supplementing the dinner with milk, salad, soup, bread, or desert "for good nutrition."

An even more recent development in food processing is the creation of *fabricated foods.* These products are not artificial, in the strictest sense of the word, because they often use conventional foods as ingredients. For example there are frozen egg substitutes, artificially colored and artificially flavored, but based on a mixture of egg whites and nonfat dry milk. According to one reviewer, when scrambled, one version of this product was "soft but acceptable" in texture, though there was "little egglike flavor."[25] Frozen breakfast ham and sausage substitutes made of textured vegetable protein, nondairy coffee creamers (some of which are high in cholesterol), and shrimp-shaped molded shrimp and bread mixtures are some other examples.[26]

It is not at the moment clear what the net energy implications of these products are. Since they involved considerable additional processing and are often frozen, they would seem to be highly energy intensive. But to the extent that some of these products substitute vegetable materials for meat, they are less energy intensive because vegetables are considerably less energy intensive than meats. There is little question, however, that those that do not make such substitutions are more energy intensive than the natural foods they attempt to replace.

Not only has there been a trend toward increasing food processing per se, but there also has been a tendency toward more energy intensive forms of processing. One example has been the large-scale switch from the sharp freezing process in which room temperatures are lowered to the quick freezing process in which the individual products are frozen in a few minutes. The latter process requires more energy to operate fans and other equpment.[27] There is a considerably more energy efficient version of "quick freezing" available (called the *Birds-eye process*), but it has failed to dominate the industry because it is associated with a greater unit labor requirement.[g] There has also been a switch in the fluid milk industry from the holding method of milk pasteurization to the much faster (30 minutes versus 16 seconds) flash pasteurization method. The latter process is somewhat more energy intensive.[28] In the prepared feeds industry, there has been a growing use of pelletizing, crimping, or flaking processes of preparing animal foods, which require more energy consumption than direct bagging.[29] Perhaps this is better classified as an example of increased

[g]A newer process, *cryogenic freezing,* involves much less energy at the food processing stage, but its total energy implications (including the production, transportation, and storage of the liquid nitrogen) are still unclear. As yet, the high cost of the liquid nitrogen has prevented widespread adoption of this technique. (See John G. Myers et al., *Energy Consumption in Manufacturing,* p. 135.)

processing, as opposed to a switch to more energy intensive processing techniques, but in any case, it has produced the same result–increased processing energy requirements.

Of course, some of the sources of excessive energy consumption discussed for industry in general in the preceding chapter apply to the food processing industries as well. Two that apply with particular force are the unnecessary dependence on natural gas and the tendency toward gross overpackaging. One of the particularly energy consumptive aspects of the latter is the use of smaller unit package sizes. In some cases, even single servings of puddings, soup, etc. are packaged in separate aluminum cans. This practice greatly increases the ratio of packaging to product and should therefore be reversed. Similar convenience can be achieved by packaging multiple servings in easily resealable containers.

Food Preparation

In Table 10–1 it was estimated that the commercial and home preparation of food was responsible for more than one-third of the energy used in the United States food system in 1970. In fact, food preparation used more than half again as much energy as did basic agricultural food production. Most of this energy was consumed directly by a handful of major food appliances: refrigerators, freezers, ovens, ranges, and dishwashers. There are, however, numerous small appliances that contribute to energy-use in food preparation. Table 10–2 contains operating data for a variety of residential appliances.

Even for a fully equipped household, the direct energy consumption due to minor appliances is on the average only about 10 to 15 percent of the appliance energy total. Clearly then, the greater potential energy savings lie in the major appliances. Nevertheless, there is something to be said for establishing a kind of energy conservation ethic that looks askance at all marginal energy-using devices and asks whether their contribution to the quality of life balances their contribution to pollution and to the depletion of limited energy and other resources. In addition, minor energy savings achieved on a wide scale can add up to substantial gains. For example, a saving equivalent to the average annual energy consumption of an electric knife (0.8 kwh/year) for every resident of the United States would result in a total energy savings equivalent to nearly one-third of a million barrels of oil per year.

Refrigeration and Freezing

Refrigeration and freezing are reasonably energy intensive means of food preservation to begin with because they require continual energy usage. There are alternate means of preventing the deterioration of foods, such as drying, chemical

Table 10–2
Average Energy Use by Residential Food-Related Appliances

Appliance	Average Operating Wattage	Average Annual Use (Hours)	Average Energy Use	
			kwh/ Year	Millions of Btu/Year
Major				
Freezer (15 ft ^{3})				
Manual Defrost	341	3,505	1,195	4.08
Frostless	440	4,000	1,761	6.01
Refrigerator (12 ft ^{3})				
Manual Defrost	241	3,020	728	2.48
Frostless	321	3,790	1,217	4.15
Refrigerator/Freezer (14 ft ^{3})				
Manual Defrost	326	3,488	1,137	3.88
Frostless	615	2,974	1,829	6.24
Range[a]				
Electric	12,200	96	1,175	4.01
Elec. Self-cleaning	12,200	99	1,205	4.11
Gas	–	–	3,106[b]	10.60[b]
Microwave	1,450	131	190	0.65
Dishwasher	1,201	302	363	1.24
Minor				
Blender	300	3	0.9	0.03
Broiler	1,140	75	85	0.29
Can Opener	100	3	0.3	0.01
Coffee Maker	894	118	106	0.36
Deep Fryer	1,448	56	83	0.28
Fry Pan	1,200	135	100	0.34
Griddle	1,200	50	46	0.16
Knife	95	8	0.8	0.03
Mixer[c]	150	10	2	0.01
Roaster	1,425	72	60	0.20
Rotisserie	1,400	52	73	0.25
Slow Cooker	200	693	139	0.47
Toaster	1,100	35	39	0.13
Toaster-Oven	1,500	165	93	0.32
Trash Compactor	400	125	50	0.17
Waste Disposer	445	67	30	0.10

[a]Range includes both oven and surface units, except in the case of microwave, which is only an oven.

[b]This figure is not directly comparable to the figures for electrical appliances, because the latter do not include energy losses in the generation of electricity.

[c]Self-standing mixer.

Sources: Science Policy Research Division of the Library of Congress, *Energy Facts* (Washington: Committee on Science and Astronautics of the U.S. House of Representatives, November 1973), p. 64; Joseph Mercieca, "Appliances and Energy Conservation," *Seminar in Engineering and Industrial Economics*, Columbia University, Spring, 1975 (unpublished) pp. 48–49; Stanford Research Institute, *Patterns of Energy Consumption in the United States* (Washington: Office of Science and Technology of the Executive Office of the President, January, 1972), p. 47.

treatment (including salting), canning, and other forms of airtight and vacuum packing and irradiation. Of these, only canning and related methods are widely usable as proven safe and effective means of preservation for a significant range of foods. However, the taste of some kinds of foods preserved by canning is significantly different from that of the fresh product. In addition, the unavailability of a quick, efficient, and inexpensive means of reestablishing the vacuum makes it necessary to rely on another means for the preservation of any foods left over from cans or jars whose seal has been broken. Refrigeration and freezing are commonly used means for this purpose.

Serious research efforts into the development of safe (both in short and long term), effective, and energy efficient alternate means of food preservation might well have a significant energy payoff.[h] But for the present and near future the chilling of foods will continue to be an effective means of inhibiting the growth of pathogens in, and the chemical deterioration of, foods without adulterating them. It is therefore important to ask how the energy efficiency of refrigeration and freezing may be improved.

In general, there is only a very slight reduction in the energy requirement per cubic foot of storage capacity for comparably designed units as the size of the unit increases. For example, a 14 cubic foot capacity frost-free, two-door refrigerator-freezer unit will consume 113 kwh/year/cubic foot on the average, while a larger 18 cubic foot unit will consume 112.7 kwh/year/cubic foot.[30] Thus the unit's total energy consumption is nearly linear with its size. Accordingly, it is very important to avoid the use of oversized units. A unit that is 20 percent larger than actual requirements will consume close to 20 percent more energy, and this will be, for all practical purposes, pure waste. On the other hand, too small a unit will be overfull, resulting in inadequate air circulation around its contents and therefore inefficient operation. Insufficient food storage capacity may also result in excessive energy-wasting shopping trips.

Because freezers must maintain lower temperatures than refrigerators (0 to 15°F versus 38 to 42°F), they consume considerably more energy per cubic foot. For this reason, proper sizing of the freezer compartment is even more important than proper sizing of the refrigeration space. However, unlike refrigerator units, energy may be conserved by loading a freezer as full as possible (without blocking the air door), because the frozen food helps to retain the cold better and reduce air spillage during door openings.[31] The relative sizing of freezer and refrigerator compartments should take this differential into account.

The energy penalty associated with the convenience of the frost-free feature in refrigerators, freezers, and combination units is very large indeed. From Table 10-2 it can be seen that frost-free units consume, on the average, between roughly 50 to 65 percent more energy than equivalent manual defrost machines.

---

[h]Of course, to the extent that the distances foods must travel and the times they must be stored are reduced, less artificial preservation will be necessary.

Compromise cycle defrost refrigerator-freezers, which automatically defrost only the refrigerator and not the freezer compartments, consume roughly 15 percent less energy than full frostless units. However this is still much more than manual deforst designs.[32] There is a partial offset to the additional energy consumption of fully or partially automatic defrost units in that hot water or other energy consuming heat sources are typically required for the periodic defrosting of manual units. But this is relatively minor.

More important is the fact that failure to regularly defrost manual units will result in frost accumulations on the cooling coils of the machine which will serve as insulation, preventing the coolant from effectively absorbing heat from the interior. This will reduce the energy efficiency of the machine progressively (particularly once the frost accumulations exceed ¼ inch or so), and may in extreme cases result in manual units becoming less efficient than those with the automatic defrost feature.[33] However, there is no reason why this kind of gross neglect should necessarily occur if machines are designed to facilitate manual defrosting and if people become aware of the utility bill savings available from simplified, once-a-month performance of this procedure.

The number and positioning of doors also effects energy use. More doors imply less spillage of cold air when food is either put into or removed from the unit and thus less make-up energy to rechill the interior. Doors located on top of a unit (parallel to the floor), as in a chest-type freezer, also conserve energy since the heavier, cold air tends to stay inside the unit when the door is opened.

Because refrigerators have not been properly insulated, there has been a tendency for moisture to condense on the outside of the units during periods of high humidity. In order to prevent that from happening, electric heater strips have been placed around the door openings, a fine example of the energy-intensive approach to design that has, in the past, been dominant in the United States. In addition, these strips are typically activated continually, whether or not the ambient humidity is high enough to cause the sweating problem they are designed to cure,[i] and may be responsible, by one estimate, for up to 16 percent of the unit's energy consumption.[34] From an energy conservation viewpoint it would make far more sense to improve the quality of insulation used. This would not only deal with the condensation problem, which is, after all, a minor issue, but more importantly would also improve the energy efficiency of the unit by reducing internal heat gain (or coolness loss).

Refrigerators, freezers, and combination units should be located as far from stoves, direct sunlight, and other heat sources as possible, in a place where there is sufficient space above and behind them to allow good air circulation to the condenser. The condenser, typically located underneath or behind the unit, should be designed for sufficiently easy (but *not* casual) access, to facilitate

---

[i]Some units do provide an "economizer" switch which allows a manual shutoff of these heater strips.

the periodic removal of dust from the coils (after the unit is unplugged) in order to prevent the deterioration of energy efficiency over time.

Additional copper windings in the unit's motor would also enhance its energy efficiently significantly, though at some incremental initial cost. In 1974 a study group at M.I.T. reportedly estimated that this simple step, in combination with such other minor modifications as the use of improved insulation, might add as much as 20 percent to the first cost of the unit but would cut energy use by about 50 percent, while maintaining the frost-free feature.[35] This tradeoff, for the average unit, implies an annual rate of return on the additional required investment of some 35 to 40 percent.[36]

The minimization of refrigerator use tends to conserve energy, whether by reducing the number and duration of door openings (and hence chilled-air loss) or by allowing the use of a smaller unit. In the latter regard, the resurrection of one reasonably common architectural feature of decades gone by might be helpful. This is the vegetable cooler compartment built into the outer wall of the building (in the kitchen). This swivel compartment permitted vegetables to be cooled by the outside air when temperatures were low enough, obviating the need for refrigerating them. In general, we probably refrigerate too many things that could as well be stored in such compartments or in naturally cool basements without additional energy consumption.

Cooking

A casual glance at Table 10-2 would appear to indicate that electric ranges are far more energy efficient than gas ranges. But this is deceptive because the data presented do not take into account the energy required to generate electricity. Since electric power stations average about 33 percent efficiency, multiplying the electric range figures by three gives a more accurate approximation of the relative energy efficiencies. By this calculation, electric ranges consume 15 to 18 percent more energy on the average than gas ranges (depending on whether or not the electric ranges have the self-cleaning feature). However, even with this adjustment, microwave ovens consume less than one-fifth as much energy as either of the other types.

Still the comparison is somewhat misleading, because one minor modification of the gas range could make it far more energy efficient. Nearly all gas ranges currently in operation in the United States utilize continuously burning pilot lights to ignite surface and oven burners. Estimates of the percentage of gas consumed by pilot lights vary, but 30 to 35% seems to be a reasonable figure.[j] Based on the average annual Btu consumption of gas ranges given in

[j] A study done by the American Gas Association estimated pilot light consumption in gas ranges at 32.1 percent of total consumption by such appliances. A Brooklyn Union Gas

Table 10-2 and the number of such appliances in operation as of 1971, the lower estimate implies that pilot lights in gas ranges consume more than 125 trillion Btu/year, the annual energy equivalent of about 22 million barrels of oil.[37] If pilot lights on other gas appliances (water heaters, space heaters, and clothes dryers) are also included, this figure would be nearly four times as large.[38] Most of this is, of course, wasted energy, in the sense that its consumption is not required for the functioning of the appliance (i.e., most of it is consumed in a standby mode).[k] Therefore a redesigned appliance without pilot lights would be far more energy efficient.

Actually, it is possible to make a considerable improvement in the energy efficiency of gas ranges by simply using so-called hypodermic pilot lights in which the gas flows through a fine hollow needle. The energy consumption of these devices is estimated at only one-fifth that of the standard design.[39] However, it is a simple matter to replace pilot lights entirely with electric sparking devices and achieve still larger energy gains.

On balance then, gas ranges with hypodermic or electric lighting devices consume 60 to 65 percent less energy than do electric ranges. The efficiency of both these designs could be significantly improved by using improved insulation. However, microwave ovens would still be far more energy efficient than either of the other major cooking appliances. It would therefore be wise energy policy to expedite the clarification of the safety questions currently surrounding microwave ovens. If clearly safe microwave ovens can be designed, they should play a major role in the conservation of energy in the United States food system.

The design of pots and pans also influences energy consumption. If the bottoms of these vessels were designed to mate more completely with surface cooking units, much of the heat that now escapes to the atmosphere could be captured for the cooking process. Furthermore, the materials of which pots and pans are made can vary the energy required for cooking. For example, baking with glass or ceramic vessels allows a reduction in temperature of about 25°F below that which would be required with metal pots and pans.[40] A fundamental reevaluation of the energy implications of alternative cooking vessel designs might well be appropriate.

A reassessment of cooking practices in the United States may also be in

---

Company estimate put the figure at about 35 percent. (Seungsoo Lee, "Effects of Pilot Lights in Gas Ranges," Seminar in Engineering and Industrial Economics, Columbia University, Spring 1974, unpublished, pp. 6-8).

[k]It has sometimes been argued that the energy consumed by pilot lights is not really wasted, since it performs some space heating function. In order for this heating function to be effective, kitchens would have to be located so that the pilot light heat would be roughly evenly distributed through the house. This is not normally the case. Typically, even during the part of the year in which this heat gain could potentially perform a useful heating function, it only serves to overheat the kitchen. This naturally adds to the load on refrigerators and freezers located there and may contribute to thermostatic unbalancing, unless the residence has a multizone thermostat system.

order. In the opinion of some authorities, people in this country tend to overcook much of their food.[41] (This seems almost indisputably true of vegetables). Certainly some other cuisines, such as Chinese and Japanese, include more quick-cooked and raw foods. The latter may not be a good idea until the use of harmful pesticides and other environmental pollutants has been substantially reduced, but energy-saving quick-cooking techniques could easily become a larger part of U.S. culinary practice. So too could the more complete utilization of foods.

Small appliances, such as electric rotisseries and broilers, can actually contribute to the conservation of energy when they are used instead of the full oven to prepare small amounts of foods. These would also benefit from added insulation. However, certain kitchen appliances, such as electric carving knives and can openers, make absolutely no sense. They consume energy while functioning no better, in fact, often worse, than their nonpowered counterparts (e.g., well-sharpened knives and manual operated wall-mounted can openers). Other kitchen appliances, though effective, are of little real value. Examples of these include orange juicers, electric coffee grinders, ice cream freezers, knife sharpeners, and ice crushers. It would make energy sense to sharply curtail or eliminate the use of powered small appliances falling into either of the last two categories, and doing so would hardly deal the standard of living a crushing blow.

Cleaning Up

As has happened so many times before, yesterday's luxury becomes today's convenience and tomorrow's necessity. So it has been with dishwashers. During the decade of the 1960s, the number of dishwashers in use more than quadrupled.[42] By 1971 the percentage of United States households owning a dishwasher was approaching 20 percent.[43] On the assumption that the elimination of dishwashers would constitute an unacceptable reduction in the standard of lving, we focus on reducing the energy consumption associated with their operation.

With dishwashers, as with so many other classes of machinery, full-capacity utilization is a key element in reducing the energy required per unit processed. Dishwashers should therefore be stacked properly and never run at part load. Considerable energy can also be saved by shutting off the machine after the washing cycle, opening the door, and allowing the dishes to dry in the room air, thus eliminating the unnecessary dishwasher dry cycle. During most of the year the heat and moisture this will add to the room will tend to increase comfort. Of course, this will be simplified by the production of dishwashers specfically designed for this sort of energy conserving air drying.

A major component of the energy consumed in the washing of dishes, whether or not is is performed automatically, is embodied in the hot water used. It would therefore be worthwhile to develop cold water dishwasher detergents. There is no reason to believe this will be difficult to do. Furthermore, presoaking dishes, pots, and pans in soapy cold water (and disinfectant) will loosen or remove a remarkable percentage of the food waste, reducing the energy required to complete the process. Besides reducing the amount of cleaning needed, soaking will also reduce the energy-wasting clogging of the dishwasher filter screen, thus producing an additional energy savings. Soaking would be facilitated by the resurrection of the double sink, made obsolete by the development of automatic dish and clothes washers.

Detergents that produce excessive suds should not be used in dishwashers, nor should excessive amounts of detergent be used. Too many suds will decrease the dishwasher's energy and performance efficiency.

Life-Style Changes

Meats constitute a large fraction of the typical diet in this country. The fact that we eat so high on the food chain is a major source of energy inefficiency in the United States food system. By way of illustration, a cornfield converts only about 1 percent of incident solar energy into edible parts of the plant. Beef cattle fed the corn will convert about 10 percent of its energy into body tissue. A human eating the beef will use about 10 percent of the stored energy. Consequently, at best, a human feeding this high on the food chain will derive about 0.0001 of the energy contained in the sunlight incident on the corn plant.[44] Eating one step lower on the food chain will therefore increase energy efficiency by a factor of 10.

This analysis is clearly oversimplified and approximate. Nevertheless, it is clear that increasing the fraction of plants and animals lower on the food chain in the nation's diet would certainly conserve a great deal of energy. Many Americans have unfortunately (from an energy standpoint) been brought up to believe that only meat and milk are real foods and that vegetables are mainly an unpleasant (with the exception of potatoes) but nutritionally necessary side dish. This attitude toward food is more prevalent here than in virtually any other nation in the world. If it were possible to alter this attitude in the direction suggested, it would certainly be an effective means of conserving energy.

We also tend to discard many edible parts of both plants and animals—even those that have already been cooked.[45] Cooking juices, rich in nutrients, are frequently thrown away. In general, a less wasteful attitude toward food would produce a considerable savings in energy along with a reduction in the generation of solid waste.

One other common energy-wasting food habit in the United States is the institution known as the private car, single-family food shopping trip. This is far less energy efficient than truck delivery of foods ordered by telephone

Yet the shopping trip has certain beneficial social and psychological aspects that would be lacking in the alternate truck delivery system. Therefore, the best compromise would seem to be the encouragement of carpooled shopping.

Summary and Conclusions

The food system in the United States employs at least 20 percent of the labor force and accounts for nearly 13 percent of the nation's total annual energy consumption. It is a complex system consisting of agriculture, agricultural support industries, food processing, food distribution, commercial and residential food preparation, food-related appliance manufacture, and a segment of the transportation industry. Only about one-quarter of the energy consumed by the food sector is directly agricultural, the remainder being roughly evenly split between food processing and transportation, on the one hand, and food preparation, on the other.

The green revolution in United States agriculture has produced enormous improvements in the yield of both animal and plant crops but only at the expense of very large increases in energy inputs. As a result the energy efficiency of United States agriculture has been in serious long-term decline. The highly energy intensive nature of the agricultural food technologies employed in the United States has made their export to most third world nations impractical. Hopes that the gains achieved by United States agriculture could be replicated throughout the world as a solution to the food problem have faded as the world energy problem came more sharply into focus. If petroleum were the only source of energy for agriculture, and agriculture were the only use of petroleum, widespread adoption of green revolution techniques would completely exhaust all of the world's known petroleum reserves in less than three decades.

Selective breeding has been widely used to maximize the ability of plants to absorb and utilize fertilizer. To the extent that artificial petroleum-based fertilizers are used, this type of genetic plant design requires heavy energy inputs. Furthermore, varieties of plants have been developed with increased ability to withstand mechanical handling and decreased taste. The benefits, such as they are, from the culture of these plants cannot be achieved without a high degree of mechanization and hence energy consumption.

It would seem that instead of designing plants for increased dependence on energy inputs, one could achieve energy gains by focusing on the development of food crop varieties with enhanced disease and pest resistance, water utilization efficiency, increased efficiency in utilizing incident sunlight, and other energy input reducing characteristics. Since only 50 of the 80,000 edible species of plants are widely cultured, there should be plenty of genetic material with which to work.

Large energy savings could be produced, along with substantial reductions in the animal feedlot pollution problem, if feedlot animal manure were widely used in place of artificial fertilizer. The recycling of only 20 percent of the manure generated in feedlots as natural fertilizer for corn agriculture, for example, would save the energy equivalent of 13 million barrels of crude oil per year. Even greater energy savings are possible by substituting so-called green manure for chemical fertilizers. This is done either by full-scale rotation of the main crop with a legume or by planting and plowing under of the legume during the off-season of the main crop. The annual savings for comparable acreage to that covered in the livestock manure example would be the energy equivalent of nearly 18 million barrels of crude oil. But, of course, the supply of green manure is neither limited nor subject to the same logistical problems as feedlot manure, so the ctual potential energy savings are far greater. In addition, there is some recent evidence that organic farms, using natural fertilizers, are not only economically viable, but produce comparable yields to conventional farms with far lower energy cost.

Agricultural machinery should be scaled so as to allow high-capacity utilization and hence energy efficient operation. The substitution of some more labor intensive techniques, such as hand spraying of herbicides, can conserve large relative amounts of energy without sacrificing performance efficiency. Highly energy intensive mechanical irrigation of farmland may be replaceable, to some extent, by a modernized version of a biblical desert farming technique, known as runoff agriculture, whose chief energy inputs are gravitational and solar.

There are considerable nutritional, taste, and energy benefits to be derived from a reversal of the current practice of excessive processing of foods. Even fresh foods rarely reach us without having first undergone everything from long periods of atmospherically controlled storage to gas ripening. Frozen prepared foods require almost continuous energy consumption and, in some cases, have questionable nutritional value. Yet they are widely consumed because they are somewhat more convenient. We are now even producing highly processed fabricated foods.

Food preparation consumes half-again as much energy as food production. Most of this energy is consumed by the major appliances: refrigerators, freezers, ranges, and dishwashers. Proper sizing of refrigerators and freezers is critical to the energy efficiency with which foods are chilled, because their energy consumption is roughly linear with storage capacity. Elimination of the frost-free feature in these machines would by itself reduce their energy consumption by 50 to 65 percent. On the other hand, improved insulation along with additional copper windings in the compressor motor of these units would save an estimated 50 percent of their present energy consumption, even without eliminating frost-free operation.

Replacing the pilot lights on gas ranges with electric ignition devices would reduce their energy usage by some 30 percent. Nationally, this simple design modification could save as much as the equivalent of 22 million barrels of oil per year. In total, modified gas ranges would consume only about 35 to 40 percent as much energy as electric ranges. The energy efficiency of both these would be enhanced by increased insulation. However, microwave ovens would still be far more energy efficient than gas or electric ranges. If safety questions can be adequately resolved, widespread adoption of microwave ovens could play a major role in reducing cooking energy requirements.

Some small cooking appliances can be energy conservative if they are used in place of large-capacity ovens when small quantities of food must be prepared. The services of many of the marginal powered kitchen appliances (e.g., electric can openers and electric knives) should be dispensed with entirely, in the service of establishing a clear-cut bias in favor of efficient use of our resources.

The energy required for cleaning dishes, pots, and pans could be reduced by the resurrection of the once common practice of presoaking and the development of cold water dishwashing detergents. If automatic dishwashers are to be used, they should be operated only at full load, and the dishes should be allowed to air dry naturally rather than going through an energy-consuming, unnecessary drying cycle.

If dietary habits in the United States were altered to increase the relative proportion of vegetables in the diet, a considerable energy saving could be achieved. This would be further enhanced by more complete utilization of the foods we consume.

It is no simple matter to estimate the total energy-savings potential of the series of proposals discussed here. Any estimate with a pretense at accuracy would have to be based on a much more detailed empirical analysis. It is interesting to note, however, that Pimentel et al., in the context of their detailed analysis of corn production, estimated that energy inputs could be halved without reducing yields.[46] Considering that, along with the estimate that refrigerator energy consumption could also be halved by design modifications and that of gas ranges reduced by on the order of 30 percent just by eliminating pilot lights, it does not seem unreasonable to expect that total energy consumption in the United States food system could be reduced by 25 to 50 percent without reducing the quality or quantity of our food by a serious effort at energy conservation.

11 The Conservation Response: An Overview

We have been told again and again that the continued expansion of energy-use is a necessary correlate of technical progress and an absolute prerequisite to the maintenance and improvement of the material standard of living. If we do not wish to descend into economic chaos or turn back the march of civilization, the only conceivable response to the energy problem is to expand the supply of energy resources to meet an essentially unrestricted increase in demand. We must take oil, coal, and gas from wherever we can find them. We must plunge headlong into the expansion of nuclear power facilities. And if in the process we pollute the air, spill oil into the seas, unbalance the ecology of wilderness areas, strip parts of the land bare of vegetation, spread the technology and capability for nuclear weapons manufacture around the globe, and create radioactive poisons that will take millenia to decay, that is part of the price we must pay. It is that or stagnation.

We have tried to show that there is another feasible response to the challenge posed by the energy problem—the conservation response. We have explored the potential for energy conservation across a variety of energy-using sectors and have found it to be very large indeed. In many cases, considerable energy could be saved by relatively simple and cheap procedures. In other cases, the procedures were somewhat more complex and expensive but still both technically feasible and economically viable. Furthermore, virtually without exception the same strategies that were found to be effective in conserving energy were also effective in reducing environmental pollution. Thus it is clear that rather than having to accept increased pollution as a cost of solving the energy problem, both the energy and pollution problems are susceptible of simultaneous solution via the conservation response.

Because the empirical work in this area has been so tentative and piecemeal to date, and because the energy conservation strategies discussed interact in complex ways, it is very difficult to give a quantitative estimate of the overall reduction in energy consumption that is possible without reducing the material standard of living. It does not, however, seem likely that the savings resulting from a serious, sustained conservation effort could be less than 30 percent, even in the medium term, and they are probably closer to 50 percent. In the long term, the figure would be larger still, but it would be difficult even to "guestimate" how large.

In order to better understand the meaning of these numbers, if total U.S. energy consumption in 1975 had been reduced by 30 percent, it would have been equal to actual consumption in 1965; had if the 1975 figure had been cut in half, it would have meant that energy consumption in that year was roughly the same as in 1955. Looked at differently, applying a 30 percent savings to the Department of the Interior projections of total United States energy use (contained in the *1974 Statistical Abstract of the U.S.*) would make the figure for 1985 equal to that of 1975; applying the 50 percent reduction would make the projections for 1995 and 1975 equivalent. In a sense, then, we are talking about offetting the effects of 10 to 20 years growth in *total* (*not* per capita) energy consumption.

It could be argued, as in fact it often has been, that even if energy conservation measures were effective in producing large reductions in energy consumption, they would at best only postpone the need for a real solution to the energy problem. And it is certainly true that as long as we continue to burn up finite fuel resources, we must eventually run out of them. But fortunately there is a permanent solution to the energy problem waiting around the corner. If we use the time that energy conservation can buy us wisely, we can fully develop it well before we even come close to running out of fossil fuel resources, without mortgaging our future and that of future generations to the nuclear genie.

Solar energy, directly in the form of sunlight and indirectly in the form of windpower, ocean thermal gradients, and the like is a virtually pollution-free, abundant energy resource that is for all practical purposes permanently renewable. It is much more equitably distributed among the nations of the world than any finite fuel resource, and it cannot be monopolized or otherwise restricted or embargoed. Much use can already be made of solar energy, but much more is possible if we will invest the resources necessary to fully develop the technology.

Whether or not we switch our energy resource development efforts to ecologically rational resources like solar energy, it is still sensible to exploit the potential for energy conservation as fully as possible. It will at minimum stretch existing supplies to the limit, reduce the pressure for expansion of energy resources sufficiently to allow more reasoned and complete evaluation of the monetary, ecological, and safety costs of such expansion, and contribute importantly to a major reduction in air, water, and solid-waste pollution. And it will do so without significantly reducing the standard of living. It is hard to ask more than that.

Notes

Notes

Chapter 1.
Energy Conservation: the Possibilities, the Limits, the Benefits

1. "(Crankcase) Oil on the Waters", *Scientific American* (February, 1973).

2. G. Tyler Miller, Jr., *Energy and Environment: Four Energy Crises* (Belmont: Wadsworth Publishing Company, Inc., 1975), p. 10.

3. *Ibid.*, p. 55.

4. Carol E. Steinhart and John S. Steinhart, *Energy: Sources, Use, and Role in Human Affairs* (North Scituate: Duxbury Press, 1974), p. 264.

5. *Ibid.*, p. 259.

6. *Ibid.*, p. 269.

7. J. Holdren and P. Herrera, *Energy* (San Francisco: Sierra Club, 1971), p. 145.

8. Bruce M. Hannon, "Bottles, Cans, and Energy," *Environment* (March, 1972).

9. William E. Franklin, et.al., "Potential Energy Conservation from Recycling Metals in Urban Solid Wastes," in: *The Energy Conservation Papers,* Robert H. Williams, ed., (Cambridge: Ballinger Publishing Company, 1975), p. 192.

10. Stanford Research Institute, *Patterns of Energy Consumption in the United States* (Washington: Office of Science and Technology, Executive Office of the President, 1972), p. 125.

11. Freeman J. Dyson, "What is Heat?" *Scientific American* (September, 1954).

Chapter 2
Building Design and Energy Consumption

1. National Bureau of Standards, *Technical Options for Energy Conservation in Buildings* (Washington: U.S. Department of Commerce, July, 1973), pp. x–xi.

2. Richard G. Stein, FAIA, "The Architecture-Energy Interface," lecture delivered to the Seminar in Engineering and Industrial Economics. Department of Industrial and Management Engineering, Columbia University, New York, February 18, 1975.

3. Richard G. Stein and Carl Stein, *Research, Design, Construction, and*

Evaluation of a Low Energy Utilization School (Washington: National Science Foundation, August, 1974).

4. Charles W. Lawrence, "Energy Use Patterns in Large Commercial Buildings," delivered to Conference on Energy Conservation: Its implications for Building Design & Operation, University of Minnesota, Bloomington, May 23, 1973.

5. Richard G. Stein, *op. cit.*

6. *Ibid.*

7. Richard G. Stein and Carl Stein, *op. cit.*

8. Richard G. Stein, *op. cit.*

9. Much of the information contained in this section is drawn from James M. Fitch and Daniel F. Branch, "Primitive Architecture and Climate," *Scientific American* (December, 1960).

10. United Nations, *Climate and Housing Design* (New York: United Nations, 1971).

11. *Ibid.*, p. 21.

12. James M. Fitch and Daniel P. Branch, *op. cit.*, p. 139.

13. United Nations, *op. cit.*, p. 20.

14. James M. Fitch and Daniel F. Branch, *op. cit.*, p. 140.

15. United Nations, *op. cit.*, p. 17.

16. James M. Fitch and Daniel F. Branch, *op. cit.*, p. 141.

17. H.R. Hay, "Energy, Technology, and Solarchitecture," *Mechanical Engineering* (November, 1973), p. 19.

18. Richard G. Stein and Carl Stein, *op. cit.*, pp. d=1/1 and d=1/2.

19. *Ibid.*, p. d=1/2.

20. Richard G. Stein, "A Matter of Design," *Environment* (October, 1972), p. 18.

21. National Board of Fire Underwriters, 1955. National Building Code.

22. Richard G. Stein, 1972, *op. cit.*, p. 18.

23. *Ibid.*

24. *Ibid.*

25. Richard G. Stein and Carl Stein, *op. cit.*, p. b/5.

Chapter 3
Heating, Cooling, and Ventilation

1. Stanford Research Institute, *Patterns of Energy Consumption in the United States* (Washington: Office of Science and Technology, Executive Office of the President, January 1972), p. 6.

2. National Bureau of Standards, *Technical Options for Energy Conservation in Buildings* (Washington: U.S. Department of Commerce, July, 1973), p. 77.

3. American Boiler Manufacturers Association, Newark, New Jersey, 1974.

4. National Bureau of Standards, *op. cit.,* p. 79.

5. *Ibid.,* pp. 85-86.

6. *Ibid.,* p. 87.

7. American Society of Heating, Refrigerating, and Air Conditioning Engineers, *Handbook of Fundamentals* (New York: 1972).

8. Richard G. Stein and Carl Stein, *Research, Design Construction, and Evaluation of a Low Energy Utilization School.* (Washington: National Science Foundation, August, 1974), p. d=5/12.

9. National Bureau of Standards, *op. cit.,* p. 93.

10. H.B. Nottage and K.S. Park, "Performance of Ventilated Luminaire," *Ashrae Transactions,* Paper No. 2112 (July, 1969).

11. T.L. Ballman, et al., "Air-Cooled Luminaires," *IES Approved Method for Photometric and Thermal Testing of Air and Liquid Cooled Heat Transfer Luminaires, LM-30* (New York, September, 1968).

12. Illuminating Engineering Society, *Illuminating Engineering* (May, 1972).

13. D.I. Musa, "Heat Recovery from Water Cooled Luminaires," Seminar in Engineering and Industrial Economics, Columbia University, Spring, 1974 (unpublished), pp. 28-29.

14. *Ibid.,* pp. 3 and 28-29.

15. Calculated from data supplied in National Institute for Occupational Safety and Health, *The Industrial Environment–Its Evaluation and Control* (Washington: Public Health Service, U.S. Department of Health, Education, and Welfare, 1973), p. 400.

16. Anthony Damas, "Reduction of Energy Consumption in H.V.A.C. Systems of Office and Industrial Buildings," Seminar in Engineering and Industrial Economics, Columbia University, Spring, 1974 (unpublished), p. 5.

17. J.R. Schreiner, "Design and Operating Cycle: Three Phase Units," *Heat Pumps–Improved Design and Performance* (New York: American Society of Heating, Refrigerating, and Air Conditioning Engineers, 1971), p. 19.

18. *Ibid.*

19. H.R. Hay and J.I. Yellott, "A Naturally Air-Conditioned Building," *Mechanical Engineering* (January, 1970).

20. *Ibid.,* p. 23.

21. "Office Planners Relying More on Solar Energy," *New York Times,* Section 8, Sunday, September 15, 1974.

22. Igal Bloch, "Solar Water Heaters with Natural Circulation for Use in Residential Homes," Seminar in Engineering and Industrial Economics, Columbia University, Spring, 1974 (unpublished), p. 11.

23. *Ibid.*, p. 2.

24. I.E. Lansberg, "Solar Radiation at the Earth's Surface," *Solar Energy*, 1967.

25. Richard G. Stein and Carl Stein, *op. cit.*, p. d=4/5.

26. *Ibid.*, p. d=7/4.

27. National Bureau of Standards, *op. cit.*, p. 27.

28. *Ibid.*, pp. 27-30.

29. *Ibid.*, pp. 31-32.

30. Lisa Hammel, "Yes, Cold-Water Wash Can Do the Job," *New York Times*, October 10, 1974.

31. National Bureau of Standards, *op. cit.*, pp. 102-103.

32. Thomas P. Bligh and Richard Hamburger, "Conservation of Energy by Use of Underground Space," *Legal, Economic, and Energy Considerations in the Use of Underground Space* (Washington: National Academy of Sciences, 1974), p. 108.

33. Carter B. Horsley, "Earth's Heat to Warm a Connecticut House," *New York Times*, January 12, 1975.

34. National Bureau of Standards, *op. cit.*, p. 138.

35. Richard G. Stein and Carl Stein, *op. cit.*, p. d=9/3.

36. *Ibid.*, p. d=9/2.

37. William W. Caudill, Frank D. Lawyer, and Thomas A. Bullock, *A Bucket of Oil* (Boston: Cahners Books, 1974), p. 54.

38. *Ibid.*, p. 50.

39. *Ibid.*, pp. 54-55.

40. R.G. Nevins and F.H. Rohles, "The Nature of Thermal Comfort for Sedentary Man," *Ashrae Transactions* Vol. 77 (Part 1, 1971).

41 P.O. Fanger, *Thermal Comfort* (Copenhagen: Danish Technical Press, 1971).

42. National Bureau of Standards, *op. cit.*, p. 115.

Chapter 4
Lighting

1. Stanford Research Institute, *Patterns of Energy Consumption in the United States* (Washington: Office of Science and Technology, Executive Office of the President, 1972), p. 7.

2. Richard G. Stein, "A Matter of Design," *Environment* (October 1972), p. 19.

3. National Bureau of Standards, *Technical Options for Energy Conservation in Buildings* (Washington: U.S. Department of Commerce, July, 1973), p. 93.

4. John Appel and James J. Mackenzie, "How Much Light Do We Really Need," *Bulletin of the Atomic Scientists* (December, 1974), p. 19.

5. Berton C. Cooper, "A Statistical Look at the Future of Lighting," *Lighting Design and Application* (July, 1971).

6. Richard G. Stein, *op. cit.,* p. 20.

7. Richard G. Stein and Carl Stein, *Research, Design, Construction, and Evaluation of a Low Energy Utilization School* (Washington: National Science Foundation, August, 1974), pp. h=1/1 and h=1/2.

8. H. Richard Blackwell, "Development and Use of a Quantitative Method for Specification of Interior Illumination Levels on the Basis of Performance Data," *Illuminating Engineering* (June, 1959).

9. John Appel and James MacKenzie, *op. cit.,* pp. 20-24.

10. R.J. Lythgoe, *The Measurement of Visual Acuity* (London: His Majesty's Stationery Office, 1932).

11. Miles A. Tinker, *Bases for Effective Reading* (Minneapolis: University of Minnesota, 1965), p. 226.

12. *Ibid.,* p. 216.

13. Richard G. Stein and Carl Stein, *op cit.,* p. h=1/3.

14. *Ibid.,* p. h=1/6.

15. F.A. Geldard, *The Human Senses* (New York: John Wiley & Sons, 1953).

16. Miles A. Tinker, *op. cit.,* pp. 208-209.

17. *Ibid.,* p. 217-218.

18. Ernest Gardner, *Fundamentals of Neurology* (Philadelphia: W.B. Saunders, 1969), p. 200.

19. Richard G. Stein, *op. cit.,* p. 20.

20. The discussion of the three types of electric lamp design that follows is largely based on: Westinghouse Electric Corporation, *Lighting Handbook* (Bloomfield, New Jersey: Westinghouse Electric Corporation, January, 1974), Chapter 3.

21. Gerald Oriol, "Lighting Energy Optimization in Modern U.S. Apartment Buildings," Seminar in Engineering and Industrial Economics, Columbia University, Spring, 1974 (unpublished).

22. *Ibid.*

23. Richard G. Stein, *op. cit.,* p. 20.

24. Westinghouse Electric Corporation, *op. cit.,* p. 3-40.

25. L.N. Weston, "Energy Conservation in Office Lighting." Seminar in Engineering and Industrial Economics, Columbia University, Spring, 1974 (unpublished).

26. *Ibid.,* p. 50.

27. Richard G. Stein, *op. cit.,* p. 25.

Chapter 5
Transportation and Energy

1. Bureau of Mines, U.S. Department of the Interior News Release, March 9, 1971.

2. Eric Hirst, *Energy Consumption for Transportation in the U.S.* (Oak Ridge: Oak Ridge National Laboratory, March, 1972), p. 27.

3. "New Amtrack Train to Chicago Slated from Here Oct. 31," *New York Times,* October 12, 1975.

4. Bureau of Railway Economics, *Yearbook of Railway Facts* (Washington: Association of American Railroads, 1974), p. 36.

5. Eric Hirst, "Transportation Energy Use and Conservation Potential," *Bulletin of the Atomic Scientists* (November, 1973), p. 40.

6. Roman Krzyczkowski and Suzanne Henneman, *Reducing the Need to Travel* (Washington: Urban Mass Transportation Administration, March, 1974).

7. *Ibid.,* p. 17.

8. Eric Hirst (1973), *op. cit.,* p. 37.

9. Eric Hirst (1972), *op. cit.,* p. 14.

10. Communications Study Group, Joint Unit for Planning Research, University College London, *Telecommunications/Transportation Substitution* (Washington: U.S. Department of Transportation, October, 1973).

11. *Ibid.,* pp. 97-101.

12. New Rural Society Project, *Quarterly Task Report #2* (Washington: U.S. Department of Housing and Urban Development, August, 1972).

13. New Rural Society Project, *Suggested Outline for Third Quarter NRS Report* (Stamford: New Rural Society Project, February, 1973).

14. For a description of the operation of such systems see Office of Policy Development and Research, *Modular Integrated Utility System* (Washington: U.S. Department of Housing and Urban Development, 1974).

15. Percival Goodman, Professor Emeritus of Architecture, Columbia University, "Urban Architecture and Energy Use," lecture delivered to the Seminar in Engineering and Industrial Economics, Columbia University, Spring, 1974.

16. Alan Voorhees, "Traffic Patterns and Land Use Alternatives," *Highway Research Bulletin,* No. 347, 1962.

Chapter 6
Land Transportation: Automobiles and Trucks

1. Bureau of Railway Economics, *Yearbook of Railway Facts* (Washington: Association of American Railroads, 1974), p. 36.

2. Eric Hirst, "Transportation Energy Use and Conservation Potential," *Bulletin of the Atomic Scientists* (November, 1973), p. 37.

3. *Ibid.*

4. John B. Heywood, "Statement of Prof. John B. Heywood, Chairman of the Panel on Emission Control Systems for Spark-Ignition Internal Combustion Engines of the Committee on Motor Vehicle Emissions (National Academy of Sciences)," in: *Automotive Research and Development and Fuel Economy* (Washington: Committee on Commerce of the United States Senate, 1973), p. 134.

5. John W. Bjerklie, "Statement of John W. Bjerklie, Chairman of the Panel on Alternate Power Systems of the Committee on Motor Vehicle Emissions (National Academy of Sciences)," in: *Automotive Research and Development and Fuel Economy* (Washington: Committee on Commerce of the United States Senate, 1973), p. 128.

6. *Ibid.*

7. Wallace Chinitz, "Rotary Engines," *Scientific American* (February, 1969), p. 92.

8. John W. Bjerklie, *op. cit.,* p. 129.

9. *Ibid.,* p. 131.

10. *Ibid.,* p. 132.

11. Graham Walker, "The Stirling Engine," *Scientific American* (August, 1973), p. 84.

12. *Ibid.*

13. John W. Bjerklie, *op. cit.,* p. 131.

14. William P. Lear, "Statement by William P. Lear, Chairman of the Board, Lear Motors Corporation," in: *Automotive Research and Development and Fuel Economy* (Washington: Committee on Commerce of the United States Senate, 1973), p. 25.

15. John W. Bjerklie, *op. cit.,* p. 130.

16. *Ibid.,* pp. 130-131.

17. Robert B. Aronson and Larry L. Boulden, "Road Testing the Electrics," *Machine Design* (October 17, 1974), p. 22.

18. Robert B. Aronson and Larry L. Boulden, *op. cit.,* p. 21.

19. Clare E. Wise, "Back from Oblivion: Special Report on Electric Vehicles 1974," *Machine Design* (October 17, 1974), p. 114.

20. John W. Bjerklie, *op. cit.,* p. 130.

21. Victor Wouk, "Statement of Dr. Victor Wouk, President, Petro-Electric Motors, Ltd.," in: *Automotive Research and Development and Fuel Economy.* (Washington: Committee on Commerce of the United States Senate, 1973), pp. 330-336.

22. C.E. Scheffler and G.W. Neipoth, G.M. Corporation, "Customer Fuel Economy Estimated from Engineering Tests," Paper 650861 (Tulsa: SAE Meeting, November, 1965).

23. Reuel Shinnar, "The Effect of the Energy Crisis on the Private Car in the U.S.," *Transportation Research* (Vol. 9, 1975), p. 90.

24. *Ibid.*

25. Committee on Commerce, *Initiatives in Energy Conservation.* (Washington: United States Senate, 1973), p. 8.

26. H.C. MacDonald, Ford Motor Company, "Effect of Emission Controls on Energy Requirements," *Energy and the Automobile* (Detroit: SAE SP-383, September 1, 1973).

27. William K. Stevens, "Sales of Foreign Cars Surging in the U.S.," *New York Times* (April 21, 1975).

28. William K. Stevens, "Rush to Smaller Cars Spurs Detroit to Alter Assembly Lines," *New York Times* (December 3, 1973).

29. Study Group on Technical Aspects of Efficient Energy Utilization, "Efficient Use of Energy," *Physics Today* (August, 1975), p. 27.

30. Committee on Commerce, *op. cit.,* p. 9.

31. J.J. Corness, Chrysler Corporation, "Passenger Car Fuel Economy Characteristics on Modern Super Highways," Paper 650862 (Tulsa, SAE Meeting, November 1965).

32. Study Group on Technical Aspects of Efficient Energy Utilization, *op. cit.,* p. 26.

33. Richard F. Post and Stephen F. Post, "Flywheels," *Scientific American* (December, 1973).

34. Marcel Gres, "Statement of Marcel Gres, Group Vice President, Tracor, Inc., Austin, Texas accompanied by Charles Kraus, Transmission Consultant," in: *Automotive Research and Development and Fuel Economy.* (Washington: Committee on Commerce of the United States Senate, 1973), pp. 395-402.

35. *Ibid.,* p. 401.

36. Reuel Shinnar, *op. cit.,* p. 91.

37. *Ibid.*

38. Marcel Gres, *op. cit.,* p. 401.

Appendix 6A
Description of Alternative Engines

1. Nicholas F. Panayatou, "The Automobile: the Problem and What Can Be Done," Seminar in Engineering and Industrial Economics, Columbia University, Spring, 1974 (unpublished), pp. 66-67.

2. Much of the information in the discussion of the Wankel is drawn from Wallace Chinitz, "Rotary Engines," *Scientific American* (February, 1969).

3. Francis W. Sears, *Mechanics, Wave Motion, and Heat* (Reading: Addison-Wesley Publishing Co., 1958), pp. 592-593.

4. The discussion of the gas turbine engine is drawn largely from James Cicarelli, "Whatever Happened to the Turbine Car," *Bulletin of the Atomic Scientists* (December, 1974); and Wallace Chinitz, "Rotary Engines," *Scientific American* (February, 1969).

5. The discussion of the Stirling engine is based largely on the information contained in Graham Walker, "The Stirling Engine," *Scientific American* (August, 1973).

6. George N. Hatsopoulos, "Statement of George N. Hatsopoulos President, Thermo Electron Corp., Waltham, Mass.," in: *Automotive Research and Development and Fuel Economy* (Washington: Committee on Commerce of the United States Senate, 1973), p. 41.

7. Clare E. Wise, "Back from Oblivion: Special Report on Electric Vehicles 1974," *Machine Design* (October 17, 1974), p. 114.

8. Much of the flywheel discussion is based on Richard F. Post and Stephen F. Post, "Flywheels," *Scientific American* (December, 1973).

Chapter 7
Land Transportation: Railroads, Bicycles, Buses, and Urban Systems

1. Bureau of Railway Economics, *Yearbook of Railway Facts* (Washington: Association of American Railroads, 1975), p. 50.

2. Much of the information in this section is drawn from Rickey U. Sunamoto, "The Diesel-Electric Locomotive vs. Electrification," Seminar in Engineering and Industrial Economics, Columbia University, Fall, 1975 (unpublished).

3. Institute of Mechanical Engineers, "Critical Factors in the Application of Diesel Engines," *Proceedings* 1969-70 (Vol. 184, Part 3P), p. 26.

4. Sunamoto, *op. cit.,* p. 23.

5. *Ibid.,* p. 18.

6. *Ibid.,* p. 16.

7. *Ibid.,* p. 20.

8. Ian Yearsley, "Are High Speed Trains on the Right Track?" *New Scientist* (6 September 1973), p. 547.

9. For an interesting analysis of these and other economic effects, see Seymour Melman, *The Permanent War Economy* (New York: Simon and Shuster, 1974).

10. *New York Times* (August 6, 1969).

11. Ian Yearsley, *op. cit.,* p. 546.

12. See Table 5-1.

13. Ian Yearsley, *op. cit.,* p. 548.

14. *Ibid.*

15. *Ibid.*

16. "Japan Extends High Speed Rail Line to her Southern Island," *New York Times* (March 11, 1975).

17. Ian Yearsley, *op. cit.,* p. 548.

18. *Ibid.*

19. Bureau of Railway Economics, *op. cit.,* p. 45.

20. John B. Hopkins, *Railroads and the Environment: Estimation of Fuel Consumption in Rail Transportation* (Springfield, Virginia: National Technical Information Service, May, 1975), pp. 4-7.

21. Bureau of Railway Economics, *op. cit.,* P. 45.

22. Bureau of the Census, *Statistical Abstract of the United States* (Washington: U.S. Department of Commerce, 1968 and 1974), pp. 562 and 568.

23. "Condition of Rails Termed Unknown Despite U.S. Effort," *New York Times* (November 23, 1975).

24. S.S. Wilson, "Bicycle Technology," *Scientific American* (March, 1973), p. 90.

25. Richard A. Rice, "System Energy and Future Transportation," *Technology Review* (January, 1972), p. 37.

26. Eric Hirst, *Energy Use for Bicycling* (Oak Ridge, Tennessee: Oak Ridge National Laboratory, 1974).

27. S.S. Penner and L. Icerman, *Energy: Demands, Resources, Impact, Technology, and Policy* (Reading, Massachusetts: Addison-Wesley Publishing Company, 1974), p. 257.

28. Samuel Walters, "Back to the Bicycle," *Mechanical Engineering* (April, 1973), p. 42.

29. Isaac Gottlieb, "Bicycles and Energy Conservation," Seminar in Engineering and Industrial Economics, Columbia University, Spring, 1974 (unpublished), p. 14-15.

30. *Ibid.,* p. 28.

31. *Ibid.,* p. 27.

32. *Ibid.*

33. William P. Lear, "Statement by William P. Lear, Chairman of the Board, Lear Motors Corporation," in: *Automotive Research and Development and Fuel Economy* (Washington: Committee on Commerce of the United States Senate, 1973), p. 25.

34. G.W. Rathenau, "Statement of Dr. G.W. Rathenau, Director, Research, Philips Research Laboratories," in: *Automotive Research and Development and Fuel Economy* (Washington: Committee on Commerce of the United States Senate, 1973), p. 226.

35. Amrit B. Bakare, "Alternative Modes of Transit to Work where Rail Systems Are Nonexistent," Seminar in Engineering and Industrial Economics, Columbia University, Spring, 1974 (unpublished), pp. 9-10.

36. Ralph Blumenthal, "Illegal Jitney Vans Popular in Riverdale," *New York Times* (November 24, 1975).

37. M. Scott MacGalden, Jr. and Charles A. Davis, *Report on Priority Lane Experiment on the San Francisco-Oakland Bay Bridge* (Sacramento, California: Bridge and Transportation Agency of the State of California, April, 1973).

38. Basil Mastorakis, "Energy Reduction in the Operation of an Urban Taxicab System," Seminar in Engineering and Industrial Economics, Columbia University, Spring, 1974 (unpublished), pp. 9-10.

39. Bruce Hannon et al., "Energy Employment and Dollar Impact of Alternative Transportation Options," in: Robert H. Williams, ed., *The Energy Conservation Papers* (Cambridge, Massachusetts: Ballinger, 1975), p. 119.

40. Ralph Blumenthal, "Toronto Offers to Sell Expertise in People-Moving," *New York Times* (October 29, 1975).

41. Donald Raskin, Transportation Research Associate, Metropolitan Transportation Authority of New York State, "Subway Car Propulsion with Energy Storage," lecture delivered to a special seminar of the Department of Mechanical Engineering, Columbia University, Spring, 1975.

42. Basil Mastorakis, *op. cit.,* p. 1.

43. *Ibid.,* pp. 12-14.

44. Calculated based on data given in the paper by Basil Mastorakis, *op. cit.,* pp. 4-5.

45. H. Bernstein and C.L. Olson, *High Capacity Personal Rapid Transit* (El Segundo, California: The Aerospace Corporation, October, 1974), pp. 4–8.

46. Ralph Blumenthal, "Personal Rapid Transit Is Tested," *New York Times* (October 23, 1975).

Chapter 8
Fluid-Borne Transportation

1. Norris McWhirter and Ross McWhirter, *Guinness Book of World Records* (New York: Bantam Books, 1973), p. 111.

2. *Ibid.*

3. Roman Krzyczkowski, *Over the Water Program Design* (Washington: Urban Mass Transit Administration, December, 1971), pp. 243–244.

4. George R. Taylor, *The Transportation Revolution, 1815–1860* (New York: Harper & Row, 1951), p. 52.

5. Much of the description in this section is based on Enrique L. Agois, "Energy Conservation in Merchant Ocean-Going Vessels," Seminar in Engineering and Industrial Economics, Columbia University, Spring, 1974 (unpublished), pp. 1–8.

6. *Ibid.,* p. 8.

7. James A. Heinen, Roger N. Shane, and James M. Bielefeld, "Minimizing Fuel Consumption in Ocean Vessels," *IEEE Transactions on Industrial Electronics and Control Instrumentation.* (February, 1973), pp. 44–46.

8. E.V. Telfer, "Economic Speed Trends," *Transactions of the Society of Naval Architects and Marine Engineers* (Vol. 59, 1951), p. 217.

9. Enrique L. Agois, *op. cit.,* p. 14.

10. H.J. Lageveen-Van Kuyk, "Cost Relations of the Treatment of Ship Hulls and the Fuel Consumption of Ships," *International Shipbuilding Progress* (July, 1967), pp. 292–311.

11. E.V. Telfer, *op. cit.,* p. 217.

12. H.J. Lageveen-Van Kuyk, *op. cit.,* pp. 292–311.

13. Richard A. Rice, "System Energy and Future Transportation," *Technology Review* (January, 1972), p. 32.

14. Eric Hirst, *Energy Consumption for Transportation in the U.S.* (Oak Ridge, Tennessee: Oak Ridge National Laboratory, March, 1972), p. 19.

15. "European Airbus opens Shuttle Service between here and Carribean," *New York Times* (November 21, 1975).

16. Committee on Commerce, *Initiatives in Energy Conservation* (Washington: United States Senate, 1973), p. 15.

17. *Ibid.*

18. Richard A. Rice, *op. cit.,* p. 32.

19. Chun-Kyu Park, "Conservation of Energy in Airplane Services," Seminar in Engineering and Industrial Economics, Columbia University, Spring, 1974 (unpublished) p. 13.

20. *Ibid.,* p. 27.

21. *Ibid.,* p. 11.

22. Joseph F. Vittek, Jr., "Is There an Airship in Your Future," *Technology Review* (July/August, 1975) p. 27.

23. Paul Kemezis, "West German Groups Revive Hope for the Zeppelin's Comeback," *New York Times* (July 15, 1975).

24. Joseph F. Vittek, Jr., *op. cit.,* p. 23.

25. Paul Kemezis, *op. cit.*

26. Joseph F. Vittek, Jr., *op. cit.,* p. 23.

27. *Ibid.,* p. 25.

28. *Ibid.,* p. 27.

29. *Ibid.,* p. 25.

30. *Ibid.*

31. *Ibid.*

Chapter 9
Industrial Products and Processes

1. Bureau of the Census, *Statistical Abstract of the United States* (Washington: U.S. Department of Commerce, 1974), p. 517.

2. John G. Myers et al., *Energy Consumption in Manufacturing* (Cambridge, Massachusetts: Ballinger, 1974), p. 18.

3. Stanford Research Institute, *Patterns of Energy Consumption in the United States* (Washington: Office of Science and Technology of the Executive Office of the President, 1972), p. 7.

4. Office of Emergency Preparedness, *The Potential for Energy Conservation* (Washington: Executive Office of the President, October, 1972), p. E1.

5. John C. Bittence, "Processes that Conserve Power," *Machine Design* (April 4, 1974), p. 100.

6. *Ibid.,* p. 96.

7. *Ibid.,* p. 100.

8. These examples, all described to the author by the consulting engineer in charge, date from the period 1973-75.

9. John C. Bittence, *op. cit.,* p. 96.

10. John G. Meyers et al., *op. cit.,* p. 38.

11. *Ibid.,* p. 34.

12. *Ibid.,* pp. 37 and 51.

13. *Ibid.,* p. 36.

14. Environmental Protection Agency, EPA *Press Briefing on Solid Waste Management and Energy* (February 8, 1974), p. 13.

15. Stanford Research Institute, *op. cit.,* p. 7.

16. Cesar D. Cortes, "Reduction of Energy Consumption by the Efficient Utilization of Steam," Seminar in Engineering and Industrial Economics, Columbia University, Spring, 1974 (unpublished), p. 6.

17. *Ibid.*

18. Committee on Commerce, *Industry Efforts in Energy Conservation,* printed at the direction of Hon. Warren G. Magnuson, Chairman. (Washington: United States Senate, October 1974), p. 92.

19. *Ibid.,* p. 90.

20. The calculations for these examples were based on data given in Cesar Cortes, *op. cit.,* pp. 15-19.

21. Committee on Commerce, *op. cit.,* p. 70.

22. Charles A. Berg, "Conservation in Industry," *Science* (April 19, 1974), p. 268.

23. "Double Duty Steam Can Save Electricity, Study Finds," *New York Times,* October 10, 1975.

24. Committee on Commerce, *op. cit.,* p. 33.

25. *Ibid.,* p. 50.

26. *Ibid.,* p. 113.

27. *Ibid.,* p. 92.

28. Robert J. Siegel, "3,700 Trillion Btu's—That's a Lot of Heat," *Think* (December, 1973), p. 28.

29. Charles A. Berg, *op. cit.,* pp. 266-267.

30. *Ibid.,* p. 267.

31. Bruce M. Hannon, "Bottles, Cans, Energy," *Environment* (March, 1972).

32. Citizens Advisory Committee on Environmental Quality, *Energy in Solid Waste* (Washington: U.S. Government Printing Office, 1974), p. 9.

33. Office of Solid Waste Management Programs, *Resource Recovery and Source Reduction: First Report to Congress* (Washington: U.S. Environmental Protection Agency, 1974), pp. 5-8.

34. Citizens Advisory Committee on Environmental Quality, *op. cit.*, p. 4.

35. Envirogenics Company, *Systems Evaluation of Refuse as a Low Sulfur Fuel*, Vol. I (Springfield, Virginia: National Technical Information Service, November, 1971), p. 1-2.

36. Citizens' Advisory Committee on Environmental Quality, *op. cit.*, p. 8.

37. *Ibid.*, p. 4.

38. *Ibid.*, p. 14.

39. Dov Iliashevitch, "Municipal Solid Waste as a Source of Energy Conservation," Seminar in Engineering and Industrial Economics, Columbia University, Spring, 1975 (unpublished), p. 18.

40. *Ibid.*, pp. 26-36.

41. The data on yield of the process is drawn from S.S. Penner and L. Icerman, *Energy: Demands, Resources, Impact, Technology, and Policy.* (Reading, Massachusetts: Addison-Wesley, 1974), pp. 270-273.

42. *Ibid.*, pp. 273-275.

43. Charles A. Berg, *op. cit.*, p. 267.

44. John G. Meyers et al., *op. cit.*, pp. 432-436.

45. *Ibid.*, p. 356.

46. Allen C. Sheldon, "Energy Use and Conservation in Aluminum Production," *Energy Use and Conservation in the Metals Industry* (New York: American Institute of Mining, Metallurgical, and Petroleum Engineers, 1975), p. 1.

47. Charles A. Berg, *op. cit.*, p. 269.

48. Gene Smith, "British Expert Terms 10% Saving Easy," *New York Times* (December 5, 1973).

Chapter 10
Food

1. David Pimentel et al., "Food Production and the Energy Crisis," *Science* (November 2, 1973), p. 444.

2. John G. Myers et al., *Energy Consumption in Manufacturing* (Cambridge, Massachusetts: Ballinger, 1974), pp. 18-19.

3. David Pimentel et al., *op. cit.*, p. 444.

4. *Ibid.*, p. 445.

5. John S. Steinhart and Carol E. Steinhart, "Energy Use in the U.S. Food System," *Science* (April 19, 1974), p. 312.

6. Carol E. Steinhart and John S. Steinhart, *Energy: Sources, Use, and Role in Human Affairs* (North Scituate, Massachusetts: Duxbury Press, 1974), p. 68.

7. *Ibid.*, p. 69.

8. John S. Steinhart and Carol E. Steinhart, *op. cit.*, p. 308.

9. David Pimentel et al., *op. cit.*, p. 444.

10. John S. Steinhart and Carol E. Steinhart, *op. cit.*, p. 313.

11. David Pimentel et al., *op. cit.*, p. 446.

12. *Ibid.*

13. *Ibid.*

14. *Ibid.*, p. 447.

15. *Ibid.*

16. *Ibid.*

17. Roy Reed, "Organic Farms Found Efficient," *New York Times* (July 20, 1975).

18. The idea that such intermediate technologies should be employed in the agricultural and industrial sectors of developing nations is discussed in E.F. Schumacher, *Small Is Beautiful: Economics as if People Mattered* (New York: Harper and Row, 1973).

19. David Pimentel et al., *op. cit.*, p. 446–447.

20. *Ibid.*, p. 446.

21. David Pimentel, William Dritschilo, John Krummel, and John Kutzman, "Energy and Land Constraints in Food Protein Production," *Science* (November 21, 1975), p. 759.

22. Much of the information in the following discussion is drawn from Michael Evenari, "Desert Farmers: Ancient and Modern," *Natural History* (August-September, 1974), pp. 43–49.

23. Arthur W. Galston, "Rotten Apples and Ripe Bananas," *Natural History* (January, 1975), pp. 30–31.

24. *Ibid.*, p. 31.

25. Jean Hewitt, "Unusual Frozen Foods–New and Fabricated," *New York Times* (February 18, 1975).

26. *Ibid.*

27. John G. Myers et al., *op. cit.*, p. 134.

28. *Ibid.*, p. 112.

29. *Ibid.*, p. 138.

30. Joseph Mercieca, "Appliances and Energy Conservation," Seminar in Engineering and Industrial Economics, Columbia University, Spring, 1975 (unpublished), p. 50.

31. *Ibid.*, p. 30-31.

32. *Ibid.*, p. 50.

33. *Ibid.*, p. 30.

34. *Ibid.*, p. 29.

35. Victor K. McElheny, "Scientists Say Refrigerators, for 20% More, Could Be Made to Run for 50% Less," *New York Times* (June 19, 1974).

36. Calculation based on Victor K. McElheny, *op. cit.*

37. Seungsoo Lee, "Effects of Pilot Lights in Gas Ranges," Seminar in Engineering and Industrial Economics, Columbia University, Spring, 1974 (unpublished), p. 8.

38. *Ibid.*, p. 6.

39. *Ibid.*, p. 6.

40. Joseph Mercieca, *op. cit.*, p. 24.

41. John L. Hess, "Fuel Shortage Could Go a Long Way toward Helping Us to Eat Better," *New York Times* (December 3, 1973).

42. Stanford Research Institute, *Patterns of Energy Consumption in the United States.* (Washington: Office of Science and Technology of the Executive Office of the President, January 1972), p. 60.

43. Bureau of the Census, *Statistical Abstract of the United States* (Washington: U.S. Department of Commerce, 1974), p. 397.

44. David M. Gates, "The Flow of Energy in the Biosphere," *Scientific American* (September, 1971), p. 100.

45. John L. Hess. *op. cit.*

46. David Pimentel et al. (1973), *op. cit.*, p. 447.

Index

Index

About the Author

Lloyd J. Dumas is Associate Professor of Industrial and Management Engineering at Columbia University. He was formerly Assistant Professor of Economics at Herbert Lehman College of the City University of New York and holds the B.A. in Mathematics (1967), the M.S. in industrial engineering (1968) and the Ph.D. in economics (1972), all from Columbia University. Dr. Dumas has spoken on energy-related topics at regional and international conferences. He is author of "Payment Functions and the Productive Efficiency of Military Industrial Firms" (*Journal of Economic Issues,* 1976); "National Insecurity in the Nuclear Age" (*Bulletin of the Atomic Scientists,* 1976); "National Security and the Arms Race" in D. Carlton and C. Schaerf, eds., *International Terrorism and World Security* (1975), and "Re-education and Re-employment of Engineering and Scientific Personnel" in S. Melman, ed. *The Defense Economy* (1970).